Bakery and Confectionery Products

Bakery and Confectionery Products

Processing, Quality Assessment Packaging and Storage Techniques

Lakshmi Jagarlamudi Ph.D
Professor and University Head
Department of Food and Nutrition
Acharya N.G.Ranga Agricultural University
College of Home Science, Lam
Guntur 522 034, Andhra Pradesh

New Delhi – 110 034

NEW INDIA PUBLISHING AGENCY
101, Vikas Surya Plaza, CU Block, LSC Market
Pitam Pura, New Delhi 110 034, India
Phone: + 91 (11) 27 34 17 17 Fax: + 91 (11) 27 34 16 16
Email: info@nipabooks.com
Web: www.nipabooks.com

Feedback at feedbacks@nipabooks.com

ISBN: 978-93-87973-65-7

Composed, Designed and Printed in India

Dedicated to my beloved Parents (Late)

Dr. Narayanaswamy Jagarlaumudi

and

Mrs. Anasuya Jagarlamudi

भारतीय कृषि अनुसंधान परिषद
कृषि अनुसंधान भवन–II, पूसा, नई दिल्ली 110 012
INDIAN COUNCIL OF AGRICULTURAL RESEARCH
KRISHI ANUSANDHAN BHAVAN-II, PUSA, NEW DELHI -110 012

डॉ. नरेन्द्र सिंह राठौड़
उप महानिदेशक (कृषि शिक्षा)

Dr. Narendra Singh Rathore
Deputy Director General (Agril. Edn.)

Phone : 011-25841760 (O)
Fax : 011-25843932
E-mail : ddgedn@gmail.com
nsrdsr@gmail.com
Website : www.icar.org.in

April 24, 2019

Foreword

I am happy to know that Dr. Lakshmi Jagarlamudi, Professor & University Head. Department of Food Science & Nutrition, ANGRAU, Andhra Pradesh, has brought a very detailed, recent and much practical information related to various aspects of Bakery and Confectionery Products in a vivid and self-explanatory manner for the benefit of Undergraduate Students of Food Science Technology.

Dr. Lakshmi had been in the field of teaching and training for more than 34 years extending her remarkable service in various Colleges of the ANGRAU where subjects like Food Science, Food Technology and Food Science & Nutrition are being offered at UG level and also at PG level. Her experience in teaching has enabled her to present various chapters pertaining to bakery and confectionery processing in an easy and reproducible manner for the students of UG. The book is a very comprehensive and much practical-oriented, which would certainly be much useful for day to day reference and in completion of course-work by UG and PG students in discipline of Food Science and Technology.

I am very much pleased to congratulate Dr. Lakshmi for her sincere efforts in timely bringing out this book, which covers 70%-80% of the revised syllabus under V Deans' Committee. It can also be use by entrepreneur startups and bakers for acquiring basic knowledge about setting up of a bakery or a confectionery unit as an elementary guide. The book is also useful for all the students appearing in any competitive/entrance examinations' in the disciplines of Food Science, Food Science & Nutrition and Food Technology.

I appreciate the Author and the New India Publishing Agency for their efforts in timely bringing out an informative book on Bakery and confectionery Products that go in long way for the food and Nutritional Security of Humankind.

Rathore

(N.S. Rathore)

Acknowledgements

I wish to express my gratitude to all those people from Bakery and Confectionery Industry and Academia, who directly and indirectly helped me in bringing out this book. I thank all those learned researchers in the field whose work also contributed to the book.

I am highly grateful to Dr N S Rathore, DDG (Edu.), ICAR, New Delhi, for his encouragement and writing a foreword to the book. It has been a pleasure for me to thank Dr (Mrs) Vijya Khader, who is my guide, mentor and a constant source of inspiration.

There are many people to thank for their help and cooperation in the preparation of the book including my students especially, B. Tech (Food Technology) and Postgraduate students of Food and Nutrition. My special thanks are due to Mrs. Preeti Sagar, Teaching Associate, College of Food Technology, Bapatla, for her untiring help in preparation of the manuscript.

It has been a pleasure to work with M/S New India Publishing Agency, New Delhi, in bringing out this publication.

Finally my deepest gratitude to my husband Prof Ravuri, daughters Dr Sasya Ravuri and Ms Spandana Ravuri, and my sons-in-law Dr Naren Sandeep Dudyala and Mr Venkat Ramakrishnam Raju Penumasta for their bountiful love and unfailing emotional support in completion of the task

Lakshmi Jagarlamudi

Preface

India, being the second largest producer of food in the world, needs to be more aggressive in food processing technology for attaining food and nutritional security of its ever growing population. Bakery holds an important place in food processing industry. The Indian bakery market has witnessed a robust growth over the past several years. Changing consumption patterns, entry of international bakery chains, rising female employment and development of new product variants and flavors have been driving forces for the market growth. Now-a-days, the availability of innovative bakery products such as photo and designer cakes, multi-grain breads, fiber biscuits, eggless cakes, etc. are extensively favoring their consumption by people of all age groups in India, owing to their convenience, taste and easily digestible nature. The fast changing eating styles and demand for fortified food products of consumers further fueling up this industry to be ever dynamic and equip with technically trained personnel.

There are many National and International Institutes offering exclusive and very specially designed Certificate/ Diploma programs for startup entrepreneurs in this field. A basic course on Bakery and Confectionery Products, however, is a mandatory course incorporated in any of the undergraduate programs (B. Sc, or B. Tech.) of Food Technology, Food Processing Technology and Food Science. In either of the cases, there is a need for a comprehensive and much practical information source to refer to. Exactly at this juncture, the Author of this book, with her more than three decades of teaching and training experience in the field, has put her efforts to write the book titled "Bakery and Confectionery Products: Processing, Quality Assessment, Packaging and Storage Techniques" under well designed chapters. The book is a very comprehensive and much practical-oriented for day to day reference and to complete course-work by UG and PG students in discipline of Food Science and Technology.

Though the book is intended to serve as a text book, it is not prepared on the basis of the syllabus of any particular educational course. Therefore, different topics are treated in such a way as to provide a comprehensive knowledge of the theoretical as well as the applied aspects involved in processing of bakery and confectionery products to gain confidence in any dedicated reader to go for a startup as well in the field. It also covers information on ingredients to bakery and confectionery products, formulae and processes for bakers, equipment for bakers and confectionery units along with quality assessment and standards. Equally well, it helps those personnel connected with industries, who supply ingredients, equipment and packaging materials for bakery and confectionery units.

The Author also feels that this book is also useful for all the students appearing in any competitive/entrance examinations' in the disciplines of Food Science, Food Science & Nutrition and Food Technology. Further, the author invites critical comments and suggestions from the students, teachers and other users of the book to improve or to modify any information for better serving all the stakeholders'.

Author

Contents

Glossary

A

Aeration: The treatment of dough or batter by charging with gas to produce a volume increase.

Absorption: Taking in or reception by molecular and or physical action. Property of wheat flour that enables it to absorb liquid.

Albumen: Egg white.

Almond Paste: Almonds ground to paste with sugar.

Angel Food Cake: A type of cake made of meringue (egg whites and sugar) and flour.

Ash: The incombustible residue left after burning matter. The term is used to denote the level of bran present in Maida.

B

Bake: To cook or roast by dry heat in a closed chamber such as an oven.

Baking Powder: A chemical leavening agent composed of soda, dry acids, and corn starch (to absorb moisture), when heated, carbon dioxide is given off, to raise the batter during baking.

Batter: A homogenous mixture of ingredients with liquid to make a mass that has soft plastic character.

Bay: A well, made in a heap of flour and other dry materials to receive the liquid ingredients for mixing.

Bleached Flour: The term refers to flour that has been treated by a chemical to remove its natural colour and make it whiter.

Bloom: A whitish coating on chocolate, caused by separated cocoa butter.

Blown Sugar: Pulled sugar that is made into thin-walled, hollow shapes by being blown up like a balloon.

Bleeding: Term applied to dough that has been cut and left unsealed at the cut thus permitting the escape of leavening gas.

Blend: A mixture of several ingredients or grades of any ingredient.

Bolting: Sifting of ground grain to remove the bran and coarse particles.

Bran: A skin or outer covering of wheat grain.

Bread: The accepted term for baked foods made of flour, sugar, shortening, salt and liquid, and leavened by the action of yeast.

Buns: Small shapes of bread dough, sometimes slightly sweetened or flavored.

Bread Dough: The unbaked mass of ingredients used for making bread.

Butter Cream: Rich, uncooked frosting containing powder sugar, butter and / or other shortening and whipped to plastic condition.

Butter Sponge: Cake made from sponge cake batter to which shortening has been added.

Butterscotch: A flavor produced by the use of butter and brown sugar.

C

Cake: A product obtained by baking a leavened batter containing flour, sugar, salt, egg,milk, liquid, flavoring, shortening, and a leavening agent.

Caramelisd Sugar: Dry sugar heated with constant stirring until melted and dark in colour.

Carbohydrates: Sugar and starches derived chiefly from fruits and vegetables sources, which contain set amounts of carbon, hydrogen and oxygen.

Cardamom: Seed of a spice plant used for flavoring.

Casien: Principal nitrogenous or protein part of milk.

Clear Flour: Lower grade and higher ash content flour remaining after the patent flour as been separated.

Cinnamon: The aromatic bark of certain trees of the laurel family, ground and used as spice flavoring.

Citron: The sweetened rind of fruit.

Corn Meal: A coarse meal made by grounding corn

Cottage Cheese: The drained curd of soured or coagulated cream pressed and mixed until smooth.

Cream: The fat portion of milk: also a thickened cooked mass of sugar, egg, milk, and a thicker used for pies and other fillings.

Creaming: The process of mixing and aerating, shortening and another solid such as sugar and flour.

Crescent Rolls: Hard crusted rolls shaped into crescents, often with seeds on top.

Cripple: A misshapen, burnet or otherwise undesirable unit.

Croissant *(krwah sawn):* A flaky, buttery yeast roll shaped like a crescent and made from a rolled-in dough.

Crusting: Formation of dry crust on surface which occurs from evaporation of water from the surface.

Custard: A sweetened mixture of eggs and milk, which is baked or cooked over hot water.

Cast Sugar: Sugar that is boiled to the hard crack stage and then poured into molds to harden.

Charlotte: (1) A cold dessert made of Bavarian cream or other cream in a special mold, usually lined with ladyfingers or other sponge products. (2) A hot dessert made of cooked fruit and baked in a special mold lined with strips of bread.

Chemical Leavener: A leavener such as baking soda, baking powder, or baking ammonia, which releases gases produced by chemical reactions

Chiffon Cake: A light cake made by the chiffon method.

Chiffon Pie: A pie with a light, fluffy filling containing egg whites and, usually, gelatin.

Chocolate Liquor: Unsweetened chocolate, consisting of cocoa solids and cocoa butter.

Cocoa: The dry powder that remains after cocoa butter is pressed out of chocolate liquor.

Cocoa Butter: A white or yellowish fat found in natural chocolate

D

Danish Pastry: A flaky yeast dough having butter or shortening rolled in to it.

Daistase: An enzyme possessing the power to convert starches in to dextrin and maltose.

Divider: A machine used for cutting dough into desired size or weight. The dough is cut by volume not by weight.

Docking: Punching a number of vertical impressions in dough piece prior to baking. Docking is done so that dough expands uniformly without bulging during baking.

Dough: The mixed mass of combined ingredients for bread/rolls and biscuits, and other baked products.

Dough Conditioner: A chemical product added to improve flour in its properties to hold gas.

Dough Room: Special rooms in which bread doughs are mixed.

Dough Temperatures: Temperature of dough at different stage of processing.

Doughnuts: A cake frequently with a center hole, made of yeast-raised or baking powder dough, and fried in deep fat.

Dry Yeast: A dehydrated form of yeast. Dry yeast has a long shelf life against fresh yeast, which is perishable steamed puddings.

H

Hard Wheat: Wheat high in protein.

Hearth Bread: A bread that is baked directly on the bottom of the oven, not in a pan.

Heavy Pack: A type of canned fruit or vegetable with very little added water or juice.

High-Ratio: (a) Term referring to cakes and cake formulas mixed by a special method and containing more sugar than flour. (b) The mixing method used for these cakes. (c) Term referring to certain specially formulated ingredients used in these cakes, such as shortening.

High-Ratio Method: Please refer to, Two-Stage Method.

Hydrogenation: A process that converts liquid oils to solid fats (shortenings) by chemically bonding hydrogen to the fat molecules.

M

Malt Extract: A syrupy liquid obtained from malt mesh, a product obtained as a result of converting the starch of sugar.

Marshmallow: A white confection of meringue like consistency.

Marzipan: Almond paste used for modeling, masking and decoration.

Masking: Act of covering with icing or frosting.

Melting Point: The temperature at which a solid becomes liquid.

Meringue: A white frothy mass of beaten egg white and sugar.

Middlings: Granular particles of the endosperm of wheat made during grinding of grains in the mills.

Mocha: A flavour combination of coffee and chocolate, but predominantly that of coffee.

Moisture: Water content of a substance.

Molasses: Light to dark brown syrup obtained in making cane sugar.

Moulder: Machine that shapes dough pieces for various shapes.

Macaroon; A cookie made of eggs (usually whites) and almond paste or coconut.

Malt Syrup: A type of syrup containing maltose sugar, extracted from sprouted barley.

Marble: To partly mix two colors of cake batter or icing so that the colors are in decorative swirls.

Marzipan: A paste or confection made of almonds and sugar and often used for decorative work.

Meringue: A thick, white foam made of whipped egg whites and sugar.

N

No-Time Dough: A bread dough made with a large quantity of yeast and given no fermentation time except for a short rest after mixing.

Nougat: A mixture of caramelized sugar and almonds or other nuts, used in decorative work and as a confection and flavoring

O

Old Doughs: Yeasted doughs that have become over fermented. This produces finished baked loaf dark in crumb colour, sour in flavour, low in volume, coarse in grain and tough in texture.

One-Stage Method: A cookie mixing method in which all ingredients are added to the bowl at once.

Othello: A small (single-portion size), spherical sponge cake filled with cream and iced with fondant.

Oven Spring: The rapid rise of yeast goods in the oven due to the production and expansion of trapped gases caused by the oven heat.

P

Patent Flour: The clean flour made by grinding the choice portion of the inner portion of the wheat.

Pie: Dessert with pastry bottom, fruit or cream filling and topped with meringue, whipped cream or pastry.

Plasticity: The consistency of feel of shortening.

Proof Box: Closed box or cabinet in which pans with molded and made up dough pieces are kept for final stage of fermentation. It should have provisions for controlled temperature and humidity.

Puff Pastry: A pastry dough inter layered with butter or shortening to give flakiness. Leavened during baking by the internally generated steam.

Q

Quick Breads: Bread product baked from lean chemically leavened batter.

R

Rolls: Small breads made from yeast leavened dough sometimes called buns, may be hard or soft crusted.

S

Shortening: Fat or oil used to tenderize baked goods or to fry products. particles.

Snaps: Small biscuits that run flat during baking and become crisp on cooling.

Slack Dough: Dough that is soft and extensible, but has lost its elasticity.

Stabilizers: Commercial preparations for use in meringue, pie fillings, icing and Marsh mallows.

Steam: Vapor formed and given off from heated water.

Straight Flour: Flour containing the entire wheat berry excluding the bran and feed.

Strong Flour: One that is suitable for the production of bread of good volume and Quality

Souffle: (a) A baked dish containing whipped egg whites, which cause the dish to rise during baking. (b) A still-frozen dessert made in a souffle dish so that it resembles a baked souffle.

Sourdough: (a) A yeast-type dough made with a sponge or starter that has fermented so long that it has become very sour or acidic. (b) A bread made with such a dough.

Sponge: A batter or dough of yeast, flour, and water that is allowed to ferment and is then mixed with more flour and other ingredients to make a bread dough.

Sponge Cake: A type of cake made by whipping eggs and sugar to a foam, then folding in flour.

Scone: A biscuit like bread.

Scone Flour: A mixture of flour and baking powder that is used when very small quantities of baking powder are needed.

Short: Having a high fat content, which makes the product (such as a cookie or pastry) very crumbly and tender.

Shortbread: A crisp cookie made of butter, sugar, and flour.

Shortening: (a) Any fat used in baking to tenderize the product by shortening gluten strands. (b) A white, tasteless, solid fat that has been formulated for baking or deep-frying.

Simple Syrup: A syrup consisting of sucrose and water in varying proportions.

Soft Wheat: Wheat low in protein.

Swiss Roll: A thin sponge cake layer spread with a filling and rolled up Corn.

Sugar-Dextrose: A form of sugar made from corn and readily fermentable.

T

Tart: Small pastries with heavy fruit filling or cream

Tempering: Adjusting temperature of ingredients to a certain degree.

Texture: Describes the measure of silkiness of the interior structure of a baked product as sensed by the touch of the cut surface.

Tunneling: A condition of muffin products characterized by large, elongated holes; caused by overmixing

V

Vegetable Colour: Liquid or pastes of vegetable nature used for colouring

W

Water Absorption: water required for obtaining bread dough of desired consistency. Flours vary in ability to absorb water. This depends on the age of flour, moisture content, wheat from which it is milled, storage conditions and milling process.

Whip: A hand or mechanical beater of wire construction used to whip materials such as cream or egg whites to a frothy consistency.

Y

Yeast: A microscopic plant that reproduces by building and causes fermentation and the giving off carbon dioxide.

Young doughs: Yeast dough that is under-fermented. This produces finished yeast goods, which are light in colour, tight in grain and low in volume.

Z

Zest: The colored outer portion of the peel of citrus fruits.

1

History, Prospects and Trends in Bakery and Confectionery Industry

Bakery Industry

Baked goods have been around for thousands of years. The art of baking was developed early during the Roman Empire. It was a highly famous art as Roman citizens loved baked goods and demanded for them frequently for important occasions such as feasts and weddings etc. Due to the fame and desire that the art of baking received, around 300 BC, baking was introduced as an occupation and respectable profession for Romans. The bakers began to prepare bread at home in an oven, using mills to grind grain into the flour for their breads. The oncoming demand for baked goods vigorously continued and the first bakers' guild was established in 168 BC in Rome. This drastic appeal for baked goods promoted baking all throughout Europe and expanded into the eastern parts of Asia. Bakers started baking breads and goods at home and selling them out on the streets.

This trend became common and soon, baked products were getting sold in streets of Rome, Germany, London and many more. This resulted in a system of delivering the goods to households, as the demand for baked breads and goods significantly increased. This provoked the bakers to establish a place where people could purchase baked goods for themselves. Therefore, in Paris, the first open-air bakery of baked goods was developed and since then, bakeries became a common place to purchase delicious goods and get together around the world. By the colonial era, bakeries were commonly viewed as places to gather and socialize.[2] World War II directly affected bread industries in the UK. Baking schools closed during this time so when the war did eventually end there was an absence of skilled bakers. This resulted in new methods being developed to satisfy the world's desire for bread. Methods like: adding chemicals to dough, premixes and specialized machinery. Unfortunately these old methods of baking were almost completely eradicated when these new methods were introduced and became industrialized. The old methods were seen as unnecessary and financially unsound, during this period there were not many traditional bakeries left.

Bakery is a traditional activity and occupies an important place in food processing industry. The bakery manufacturers in India can be differentiated into the three broad segments of bread, biscuits and cake. About 1.3 million tonnes of the bakery products industry in India is in the organized sector out of 3 millions tonnes, while the balance comprises of unorganized, small-scale local manufacturers. Though, there are sufficient automatic and semi-automatic bread as well as biscuit manufacturing units in India but there are still number of people prefer fresh bread and other products from the local bakery. After entry of Pizza and Burgers' MNCs in the country, people are changing their tastes also. Today, they are not restricted to bread, cake and biscuits but to other bakery products also. The consumers are increasingly going for newer options with respect to bakery products. With the ventures of few Companies like Britannia, Biskfarm, and Morish etc. competition has increased. Also, the Indian market is witnessing the proliferation of bakery cafe chains in the form of Baristacafe, coffee day, Café Coffee & Monginis etc.

Armed with better technology, know-how and novel ideas, these foreign companies have made rapid inroads into the lucrative market within a very short span of time. Though the demand for bakery products in India has always been on the rise, there is clearly a lack of awareness and the will to break new ground, which has helped global players to gain brownie points while exploring the market in India. The unorganized sector accounts for about half of the total biscuit production estimated at 1.5 million tonnes. It also accounts for 85 per cent of the total bread production and around 90 per cent of the other bakery products estimated at 0.6 million tonnes. The last includes pastries, cakes, buns, rusks and others.

Indian Baking Industry and Prospects

The Indian bakery sector consists of some of the large food categories like breads, biscuits, cakes etc. The branded packaged segment in this sector had a size of Rs. 17,000 crore in last financial year and is expected to grow at phenomenal rate of 13–15 per cent in the next 3–4 years. Within biscuits, 3–4 large-sized players viz. Britannia, Parle, ITC, Cadburys comprise about 75 per cent of the market. The breads and cakes market is much more fragmented with multiple regional and local players. International players like United Biscuits, Unibic have gained prominence in the last few years in their specific product segments. Going ahead, the sector is expected to see some more of the international brands entering the Indian market.

The Indian bakery industry is dominated by the small-scale sector with an estimated 50,000 small and medium-size producers, along with 15 units in the organized sector. Apart from the nature of the industry, which gravitates to the markets and caters to the local tastes, the industry is widely dispersed also due to the reservation policies (relating to the small scale industries) of the government.

If we believe on the reports, the bakery industry has achieved third position in generating revenue among the processed food sector. The market size for the industry is pegged at US$ 4.7 billion in 2010 and is expected reach US$7.6 billion by 2015. It also mentions that the shining star of the sector remains the biscuits industry, which is expected to outperform the growth of the sector overall. While the figures are indeed encouraging, there is a flip side to this story. As the business and the industry thrives, the challenges accruing out of it are also growing at a fast pace. Admittedly, the Indian bakery industry is not really geared up to face the daunting task that lies ahead, which is of striking a balance.

The per capita consumption of bakery products in India is very low which is about 1–2 kg per annum, which is comparatively much lower than the developed countries where consumption is between 10 and 50 kg per annum. The growth rate of bakery products has been tremendous in both urban and rural areas. Bakery segment has increased and matured to a great extent, recently. This tremendous growth has happened due to two reasons. Firstly, due to the availability of better ingredients from chocolate, toppings, fillings, flavours etc. Secondly, education abroad has brought in many new players striving to produce products of international standard rather than products of mediocre quality. Number of players is increasing slowly. More and more people are starting to take this up as a profession from the house, after doing short/long courses.

Current Research on the Retail Bakeries Industry

Like many industries, retail bakeries see rising costs in fuel, health care, and other expenditures cut into their bottom line and increase the cost of doing business. However, there has been an uptick in demand, including trends such as cupcake stores and gluten-free baked goods.

Despite economic worries, the industry is expanding its customer base while other food service sectors continue to suffer. Consumers are now willing to spend moderately and demand high quality for their dollar; an equation for which retail bakeries are a solution. The most promising sector for the retail baking is cakes. Wedding cakes in particular are gaining ground as the economy rebounds.

Trends

Trends have manifested themselves in the various new launches/introductions that we have seen in the past few years by leading players be it Britannia's

Nutrichoice range, smaller packs of Good Day, ITC dark fantasy, Parle's Happy Happy and Parle-G Gold.

Factors for Growth

Recently, a lot of bakers have gotten into three dimensional cakes and theme cakes. Cutting off from the regularity, bakers are now looking at experimenting with many more ingredients like rice treats, and inculcating them into cake designs. Some bakers even make use of wooden planks for support. There is also something as sugar crystal sculptures, where they try and use them as per the theme of the cake. The biscuit category is expected to continue its growth trajectory of 15 per cent going ahead. Growth in bread would be relatively slower.

Challenges and Opportunities

The challenges would be category-specific. The biscuits category has seen rapid growth in the last few years. Implementation of packaging standardization norms appears to be the big challenge. Volatility in input costs is expected to remain and this would add to the woes. In bread, profitability has remained the focal point for some time. Players have been looking to increase share of value-added products while focusing on operational efficiencies linked to daily distribution. The challenge for cakes would be to expand the consumption of packaged cakes. In fact, this challenge is also a significant opportunity for this particular segment.

Regulatory Aspect

A steady rise in cost of raw materials, quality variations at source, ignorance about the new trends especially amongst the unorganized sector, stringent government regulations and legal complexities are a few of the many hurdles that cripples the bakery industry in India today. And unless, the head honchos of the industry find a way to circumnavigate these problems, if not overcome it completely, there is little hope of the bakery industry in India to sustain or to succeed.

Apart from the nature of the industry, which gravitates to the markets and caters to the local tastes, the industry is widely dispersed also due to the reservation policies (relating to the small scale industries) of the government.

Global Bakery and Pastry Industry

The global baked goods market has shown rapid recovery following the economic recession, recording strong growth over recent years. Factors fuelling market expansion include convenience, affordability and health benefits of baked goods products. Demand for healthier fortified baked products has also driven sales. The baked goods industry, encompassing products such as bread, cereals, cakes, biscuits, pastries and scones, is well established in industrialized regions such as Western Europe and North America. Product innovation and healthier products and ingredients are fuelling market growth in these regions. Emerging markets such as Latin America, Middle/East Africa and Asia represent good market potential as western lifestyles and eating habits continue to be adopted in these regions.

Biscuits and bread which are considered to be the major bakery product and they account for 82% of all bakery production. The unorganized sector accounts for about half of the total biscuit production estimated at 1.5 million tonnes. It also accounts for 85% of the total bread production and around 90% of the other bakery products estimated at 0.6 million tonnes. The last includes pastries, cakes, buns, rusks and others.

Biscuits are estimated to enjoy around 37% share by volume and 75% by share by value of the bakery industry. The organized sector caters to the medium and premium segments, which are relatively less price-sensitive. The organized sector is unable to compete at the lower price range due to the excise advantage enjoyed by the informal sector. The organized segment in biscuits has witnessed a steady growth of about 7.5%, conforming broadly to the growth rate of GDP.

Biscuits constitute about 7% of the Rs 478 billion FMCG markets in India. During 2003–04 biscuits market grew at double digit (about 11%) compared to a growth of 1.4% for the FMCG industry as a whole, and 4.4% average growth over last five years (1999–2003). In India the annual per capita consumption of branded confectionery is still under 100 gms. Hard-boiled candy is reserved for the small-scale sector. There are about 5,000 units catering to the local markets. The big players have used a mix of franchise arrangements (with small units) and product formulations to get out of the reservation mode.

The total contribution of the sugar boiled confectionery market in the organized sector, comprise plain, hard-boiled candies, toffees, éclairs and gums is around Rs. 20 billion. Add to this the unorganized sector and the market for all types of confectionery is Rs. 50 billion. However, in terms of value the organized sector commands 60% of the market share. With the exit of MNCs and other established organized players from very low priced (25 paise) category, the unorganized sector has grown very fast. MNCs and high-powered advertising support substitute products like chewing and bubble gums. With ₹3,250 million market shares, the gum and mint market is growing at a rate 10–15% annually. Fruit and mint rolls being marketed by companies with sound strategies are going ahead rapidly.

Biscuits Demand: Past and Future		Market Structure Market Segmentation	
Year	**Thousand MT**	**Segment**	**Share (%)**
2000–01	1110	Organised	50
2001–02	1188	Informal	50
2002–03	1307	North	36
2003–04	1444	East	19
2004–05	1523	West	23
2005–06	1607	South	22
2006–07	1696	**Market Growth Rates**	
2007–08	1804	1990–91, 1996–97	5.5%
2008–09	1920	1996–97, 2001–02	5.8%
2009–10	2043	2001–02, 2006–07	7.4%
2014–15	2758	2004–05, 2009–10	6.4%
		2009–10, 2014–15	6.2%

Bread Demand: Past and Future		Market Structure Market Segmentation	
Year	**Rs bn.**	**Segment**	**Share (%)**
2000–01	11.90	Organised	15
2001–02	12.85	Informal	85
2002–03	13.85	North	35
2003–04	14.80	East	10
2004–05	15.85	West	30
2005–06	16.90	South	25
2006–07	17.90	**Market Growth Rates**	
2007–08	18.95	1990–91, 1996–97	5.4%
2008–09	20.00	1996–97, 2001–02	7.9%
2009–10	21.10	2001–02, 2006–07	6.9%
2014–15	26.90	2004–05, 2009–10	6.0%
		2009–10, 2014–15	5.0%

Product Variation	
Segment	**Share (%)**
Milk Bread	85
Brown Bread	10
Fruity	3
Nutritional and other specialty	2

Source: Ministry of Food Processing Industries

Nutritional Significance

The main ingredients in baked goods are flours from cereals such as wheat, maize, and sorghum. In general, all flours contain valuable amounts of energy, protein, iron and vitamins, but the degree of milling will influence the final nutritional content.

Traditional milling produces flour which contains all of the crushed grain. Although these flours often have a coarse texture and an off-white colour, they contain many B vitamins, minerals, and also fibre. The desire for white flour however, has led to a milling process which removes the bran and, as a result, many nutrients are lost.Many baked products incorporate high levels of fat, sugar, and sometimes fruit or nuts, and this will increase the energy content of the products.

Nutritional consideration in formulation of bakery product development in bakery technology-biscuits based on composite flour, biscuits with different flavours, special biscuits vitamin fortified, high fibre, low sugar and fat biscuits. Protein in terms of quantity and quality is a vital nutrient. Wheat flour protein- 12–16% deficient in lysine essential amino acids-80 the quality of bread is below mild and meat class.

Chocolate and Confectionery Industry

History

The **history of chocolate** begins in Mesoamerica. Fermented beverages made from chocolate date back to 1900 BC. The Aztecs believed that cacao seeds were the gift of Quetzalcoatl, the god of wisdom, and the seeds once had so much value that they were used as a form of currency. Originally prepared only as a drink, chocolate was served as a bitter, frothy liquid, mixed with spices, wine, or corn puree. It was believed to have aphrodisiac powers and to give the drinker strength. Today, such drinks are also known as "Chilate" and are made by locals in the South of Mexico.

After its arrival to Europe in the sixteenth century, sugar was added to it and it became popular throughout society, first among the ruling classes and then among the common people. In the 20th century, chocolate was considered a staple, essential in the rations of United States soldiers at war.

The word "chocolate" comes from the Classical Nahuatl word *chocolātl*, and entered the English language from the Spanish language. Chocolate may be the "food of the gods," but for most of its 4,000-year history, it was actually consumed as a bitter beverage rather than as a sweet edible treat. Anthropologists have found evidence that chocolate was produced by pre-Olmec cultures living in present-day Mexico as early as 1900 B.C. The ancient Mesoamericans who first cultivated cacao plants found in the tropical rainforests of Central America fermented, roasted and ground the cacao beans into a paste that they mixed with water, vanilla, honey, chili peppers and other spices to brew a frothy chocolate drink. Olmec, Mayan and Aztec civilizations found chocolate to be an invigorating drink, mood enhancer and aphrodisiac, which led them to believe that it possessed mystical and spiritual qualities. The Mayans worshipped a god of cacao and reserved chocolate for rulers, warriors, priests and nobles at sacred ceremonies.

When the Aztecs began to dominate Mesoamerica in the 14th century, they craved cacao beans, which could not be grown in the dry highlands of central Mexico that were the heart of their civilization. The Aztecs traded with the Mayans for cocao beans, which were so coveted that they were used as currency. (In the 1500s, Aztecs could purchase a turkey hen for 100 beans.) By some accounts, the 16th-century Aztec emperor Montezuma drank three gallons of chocolate a day to increase his libido.

In the 1500s, Spanish conquistadors such as Hernán Cortés who sought gold and silver in Mexico returned instead with chocolate. Although the Spanish sweetened the bitter drink with cane sugar and cinnamon, one thing remained unchanged: chocolate was still a delectable symbol of luxury, wealth and power. Chocolate was sipped by royal lips, and only Spanish elites could afford the expensive import.

Spain managed to keep chocolate a savory secret for nearly a century, but when the daughter of Spanish King Philip III wed French King Louis XIII in 1615, she brought her love of chocolate with her to France. The popularity of chocolate quickly spread to other European courts, and aristocrats consumed it as a magic elixir with salubrious benefits. To slake their growing thirst for chocolate, European powers established colonial plantations in equatorial regions around the world to grow cacao and sugar. When diseases brought by the European explorers depleted the native Mesoamerican labor pool, African slaves were imported to work on the plantations and maintain the production of chocolate.

Chocolate remained aristocratic nectar until Dutch chemist Coenraad Johannes van Houten in 1828 invented the cocoa press, which revolutionized chocolate-making. The cocoa press could squeeze the fatty cocoa butter from roasted cacao beans, leaving behind a dry cake that could be pulverized into a fine powder that could be mixed with liquids and other ingredients, poured into molds and solidified into edible, easily digestible chocolate. The innovation by van Houten ushered in the modern era of chocolate by enabling it to be used as a Confectionery ingredient, and the resulting drop in production costs made chocolate affordable to the masses.

In 1847, British chocolate company J.S. Fry & Sons created the first solid edible chocolate bar from cocoa butter, cocoa powder and sugar. Rodolphe Lindt's 1879 invention of the Conching machine, which produced chocolate with a velvety texture and superior taste, and other advances allowed for the mass production of smooth, creamy milk chocolate on factory assembly lines. You don't need to have a sweet tooth to recognize the familiar names of the family-owned companies such as Cadbury, Mars and Hershey that ushered in a chocolate boom in the late 1800s and early 1900s that has yet to abate. Today, the average American consumes 12 lbs. of chocolate each year, and more than $75 billion worldwide is spent on chocolate annually.

Chocolate market is estimated to be around 1500 crores growing at 18–20% per annum. Cadbury is the market leader with 72% market share. The company's brands (Five star, Gems, Eclairs, Perk and Dairy milk are leaders in the segment). Till early 90's Cadbury had a market share of above 80%. Approximately about 15% market share was snatched by Nestle with introduction of Kit-Kat. The other players in the segment are GCMMF (Gujarat Co-operative Milk Marketing Federation) and CAMPCO (Central Areca nut and Cocoa Manufacturers and Processors Co-operative). Competition is on rise as chocolate giants Hershey's and Mars are trying to grab a bite of Indian chocolate market share. The per capita consumption of chocolate in India is 300 gram compared with 1.9 kilograms in developed markets such as the United Kingdom. Over 70 per cent of the consumption takes place in the urban markets. Margins in the chocolate industry range between 10 and 20 per cent, depending on the price point at which the product is placed.

Another estimate puts the figure at 25000 tonnes. The chocolate wafer market (Ultra Perk etc) is around 35 % of the total chocolate market and has been growing at around 13% annually. As per Euromonitor study, Indian candy market is currently valued at around USD 664 million, with about 70%, or USD 461 million, in sugar confectionery and the remaining 30%, or USD 203 million, in chocolate confectionery. Entire Celebrations range marketshare is 6.5%. The global chocolate market is worth $75 billion annually. The chocolate market in India has only three big players, Cadbury, Nestle and Amul. New brands such as Sweet World, Candico and Chocolatiers are present in several malls.

Mithai- the traditional Indian sweats is getting substituted by chocolates among upwardly mobile Indians. Instead of buying sweats on Raksha Bandhan, sisters prefer offering chocolates to their brothers. This is the reason for sudden spurt in advertisement between

The majority of manufacturers, including Mars International India Pvt. Ltd and Mondelez India Foods Ltd, are expanding their premium chocolate confectionery offers in India. Premium chocolate products are mostly available in modern retail channels, such as hypermarkets, supermarkets, and chemists/ druggists across India, though they are also slowly penetrating rural India as well.

Mondelez India Foods Ltd remained the largest player in chocolate confectionery in India in 2016, holding a 49% value share. While increasing penetration and the launch of smaller SKUs helped Mondelez achieve 8% value growth in chocolate confectionery in 2016, the company lost share during the year, with smaller players, outside the top eight, making notable gains.

Prospects

Chocolate confectionery value sales are expected to see a constant value CAGR of 8% over 2016–2021, to reach INR162 billion. Increasing availability in rural India and an expanded offer will contribute to chocolate confectionery growth during the forecast period.

The majority of manufacturers, including Mars International India Pvt Ltd and Mondelez India Foods Ltd, are expanding their premium chocolate confectionery offers in India. Premium chocolate products are mostly available in modern retail channels, such as hypermarkets, supermarkets, and chemists/ druggists across India, though they are also slowly penetrating rural India as well.

Competitive Landscape

Mondelez India Foods Ltd remained the largest player in chocolate confectionery in India in 2016, holding a 49% value share. While increasing penetration and the launch of smaller SKUs helped Mondelez achieve 8% value growth in chocolate confectionery in 2016, the company lost share during the year, with smaller players, outside the top eight, making notable gains.

References

Cereal Processing Technology, ed. G. Owens, Woodhead Publishing, Cambridge, 2001.

D. Manley, Technology of Biscuits, Crackers and Cookies, (3rd Edition) Woodhead Publishing, Cambridge, 2000.

Encyclopedia of Food Science and Technology, ed. F. J. Francis, volumes 1–4, John Wiley & Son, 1999.

Sultan, W.J. Practical baking. 5th edition. New York : Van Nostrand Reinhold, 1990

www@wheatfoods.org

2

Types of Bakery and Confectionery Products

Types of Bakery Products

Baked goods are produced from either doughs or batters which are a mixture of flour and water made by mixing, beating, kneading or folding. The processing method depends on the ingredients being used and the product being made. Bread is either leavened or unleavened. Leavened bread is made from a mixture of flour, yeast, salt and water. Unleavened bread does not contain yeast and therefore does not rise. It is flat bread that is quicker to make than yeast-bread.

Table 1: Classification of bakery products

Categories of Bakery Product	Types within Each Category
Unsweetened goods	Bread: sliced, crusty, par-baked, ethnic rolls; soft, crusty crumpets English muffins Croissants Pizza base raw pastry.
Sweet goods	Large cakes: Plain, Fruited Pancakes Doughnuts, Waffles Cookies Biscuits American muffins Buns Wafers.
Filled goods	Tarts: Fruit, Jam pies, Meat, Fruit sausage rolls, Pasties cakes, Cream, Custard Pizza Quiche

Source: Adapted from Smith, J.P. et.al., Crit. Rev. Food Sci. Nutr. 44, 19, 2004

Bread

Bread is a staple food prepared from a dough of flour and water, usually by baking. Throughout recorded history it has been popular around the world and is one of the oldest artificial foods, having been of importance since the dawn of agriculture. Proportions of types of flour and other ingredients vary widely, as do modes of preparation. As a result, types, shapes, sizes, and textures of breads differ around the world. Bread may be leavened by processes such as reliance on naturally occurring sourdough microbes, chemicals, industrially produced yeast, or high-pressure aeration. Some bread is cooked before it can leaven, including for traditional or religious reasons. Non-cereal ingredients such as fruits, nuts and fats may be included. Commercial bread commonly contains additives to improve flavor, texture, color, shelf life, and ease of manufacturing. Many types of breads are manufactured around the world which have been listed below

Professional bread recipes are stated using the baker's percentage notation. The amount of flour is denoted to be 100%, and the other ingredients are expressed as a percentage of that amount by weight. Measurement by weight is more accurate and consistent than measurement by volume, particularly for dry ingredients. The proportion of water to flour is the most important measurement in a bread recipe, as it affects texture and crumb the most. Hard wheat flours absorb about 62% water, while softer wheat flours absorb about 56%.Common table breads made from these doughs result in a finely textured, light bread. Most artisan bread formulas contain anywhere from 60–75% water. In yeast breads, the higher water percentages result in more CO_2 bubbles and a coarser bread crumb. One pound (450g) of flour yields a standard loaf of bread or two French loaves. Calcium propionate is commonly added by commercial bakeries to retard the growth of molds.

Soda Bread

Soda bread is prepared by substituting baking soda for yeast in a traditional bread recipe. Soda bread is very sweet with a light texture, and is frequently flavored by adding nuts or raisins to the dough.

Sourdough

Sourdough bread is baked with certain bacteria that produce lactic acid and create a sour taste. Sourdough typically has a crispy outer crust and a softer, crumbier interior.

White bread

Classic white bread has actually been around for a relatively short time, compared to other breads. It is made with bleached, chemically refined white flour, resulting in its white color. Similarly, whole wheat bread is made with whole wheat flour, which is not refined

Cracker

Crackers are like small segments of very crispy bread, originally made by combining flour, salt and water and baking the mixture. Crackers are distinguished from bread because they are not prepared with leavening. There are countless brands and flavors of crackers available today.

Crouton

A crouton is a small piece of very crunchy bread that has been baked twice, usually after bread has gone stale. Croutons are cut into small cubes, seasoned, and used to garnish foods like soups and salads.

Date Nut

Date nut bread is made by combining dates, walnuts, and sometimes pecans, with egg, baking soda and a dough-like batter. It is rather rich and sweet, and is often topped with cream cheese.

Panettone

A traditional Italian bread served at Christmas, panettone is prepared by curing dough for many days, then adding a variety of candied fruits, raisins, and sometimes lemon zest. The finished product is a tall loaf with an airy and light interior, and a sweet flavor.

Marble Bread

Marble bread is made by combining pumpernickel and rye dough, and twisting the two together to create a swirl pattern in the finished product. Marble bread is baked in dense loaves and often used for deli sandwiches.

English Muffin

The English muffin is a round yeast roll, often prepared by cooking dough on a griddle. Like a crumpet, an English muffin can be very dense and filled with air pockets. They are most often used as a breakfast roll, particularly as a base for breakfast sandwiches.

Focaccia

Focaccia bread was originally made in Italy. It tends to be relatively flat, as it is not kneaded before it is baked. It is not an entirely flat bread, because yeast is still one its ingredients, which causes it to rise slightly. Focaccia has a very rich flavor, and retains a lot of moisture, since it is brushed with olive oil before it is baked.

Fruit Bread

Fruit bread comes in almost countless varieties, consisting of dried fruit, and sometimes nuts. One of the most popular fruit breads is banana bread. Fruit bread is prepared very much like a cake, usually in a pan rather than as a freestanding loaf, and the mixture does not rise.

Matzo

Matzo is an unleavened flatbread, with a crisp and crunchy consistency similar to crackers, traditionally eaten on the Jewish holiday known as Passover.

M'smen

M'smen is traditionally made in Morocco. It is a flatbread, usually eaten as a breakfast food, with a flaky texture and a buttery flavor.

Arepa

Arepa is a bread produced in South America. It has a similar texture to a soft tortilla, but is thicker, where tortillas are flat. It is made from maize flour, and frequently used for sandwiches with meat and cheese.

Baguette

Baguettes are a very popular type of French bread, characterized by their long tube-like shape, as well as their crunchy crust and soft interior. Baguettes can be up to two feet long, and are used for a variety of purposes outside of sandwiches.

Báhn Mì

Báhn Mì is like the Vietnamese version of a baguette. It is made with a combination of rice flour and wheat, and used almost exclusively for traditional Vietnamese sandwiches. Like a baguette, its crust is very crunchy while its inside is softer.

Bagel

Perhaps one of the most popularly consumed kinds of bread, bagels are made with yeast dough. They are rolled, boiled, and baked in an oven, and they have a denser texture than other types of bread. There are countless varieties and flavors of bagel available, including blueberry, everything, onion, whole wheat, and many more.

Bialy

A bialy is a round chewy roll, somewhat like a bagel, originally made in Bialystok, Poland. Bialys have a small indent in the center, which are commonly filled with onions and poppy seeds to provide flavor before they are baked. Like bagels, bialys are made with yeast, but they are prepared differently.

Breadstick

Breadsticks are available in nearly every restaurant in nearly every country of the world. They are long, thin pieces of bread that are baked for a long time, usually until they become crisp. The extra baking time lengthens the amount of time that the bread can be kept before being eaten.

Brioche

A brioche is a glazed roll with a sweet and rich flavor. It is often served with breakfast foods because of its sweetness. It is made by combining yeast with butter and eggs, and glazing with an egg wash after baking. Brioche is sometimes flavored, particularly with almonds.

Challah

Challah is traditionally Jewish bread. It is braided before it is baked, giving it a very unique appearance. It has a sweet flavor, and is typically baked with yeast, eggs, honey, and flour.

Ciabatta

Ciabatta is an Italian loaf bread, with dense crumbs and a very hard and crisp crust. It is baked with wheat and often flavored with olive oil, rosemary or other spices, and dusted with flour when it comes out of the oven. Ciabatta is very frequently used for sandwiches, especially Panini, as it toasts particularly well,

Cornbread

Cornbread is made by baking corn that has been ground down into meal. Egg and buttermilk are often combined with the cornmeal before baking, making cornbread very cake-like in texture and taste. Cornbread can be very dense and crumby.

Croissant

Croissants are flaky, buttery, and very rich, and shaped like crescent moons. They are French rolls, made by baking puff pastry and yeast dough together in layers. Croissants are traditionally considered a breakfast pastry, and are often served with coffee in European countries, particularly France. Chocolate croissants are very popular as well; they are baked the same way, but a piece of dark chocolate is placed in the dough first.

Scone

A scone is classified as a quick bread. They are prepared by combining flour, baking soda, sugar, eggs, milk and butter and baking the mixture. The texture of a scone is very dense and dry, with a very hard crust. They are traditionally eaten as a breakfast food, with butter, clotted cream, or honey, and are often flavored with fruit in the dough, such as blueberries or raisins.

Doughnuts

A doughnut, also spelled as donut, is a type of sweet deep fried pastry. Variations on doughnuts are popular all over the world, where they are called by a wide variety of names, although the basic incarnation of fried sweet dough remains the same. The doughnut is often sprinkled with sugar or other toppings, and may be frosted or glazed as well. Associated in some nations with breakfast, the doughnut is generally agreed to be a sweet and delicious indulgence. Some countries also consume donut holes, a nod to the missing chunk of pastry in a classical doughnut. There are two primary divisions of doughnut: yeast doughnuts and cake doughnuts. The yeast doughnut hails back to the origins of doughnuts as leftover baking scraps, and tends to be lighter and fluffier with a distinctive yeasty flavor which some consumers find quite appealing.

Yeast doughnuts are sometimes baked, as this is perceived to be healthier than deep frying. Cake doughnuts are heavier, with a denser cake-like texture to them. Fried pastries come in many shapes, but the doughnut is usually only recognized as a circular pastry with a hole in the middle. Doughnuts can be filled or glazed, but the basic shape will remain the same. When cooked properly, the doughnut will retain a round shape and a crisp exterior.

Bagels

A bagel is a bread product, traditionally made of yeasted wheat dough in the form of a roughly hand-sized ring which is first boiled in water and then baked. The result is a dense, chewy, doughy interior with a browned and sometimes crisp exterior. Bagels are often topped with seeds baked onto the outer crust, with the traditional being poppy or sesame seeds. Some have salt sprinkled on the surface of the bagel.

Muffins

Muffins, usually sweet cupcake-shaped breads, may require a double explanation in the UK. Typically, and until recently, muffin in the UK referred to what most call the English muffin: a flat round yeast bread that is split and toasted. The US type is very different, and is defined as a quick bread, since it doesn't contain yeast. Many variants exist and both types are sold in the UK now. In fact the term may come from an old French word introduced into English, moufflet, in approximately 1000 CE. The term may have once been mufflins, and at least was so used by some Irish immigrants to the US in the early 20th century. Muffins get their characteristic rise from baking powder, or sometimes baking soda, instead of yeast.

Puffs

Cream puffs are always made with the same method. Water and butter are brought to a boil in a saucepan, then flour is stirred in all at once. The dough that forms is stirred and cooked for a few minutes, until it leaves the sides of the pan. Then, off the heat, eggs are beaten in, one at a time, until the dough is shiny, glossy, and sticky. The dough is dropped onto cookie sheets and baked in a hot oven. Steam forms as the puffs bake, and the strong gluten structure formed by beating the dough stretches to hold the steam, then sets into place as the heat coagulates (sets) the protein. The puffs will be dark golden brown, with a hollow center crisscrossed with soft filaments of dough.

Rusk

These types of bread are double baked. Rusk are popular teatime snacks .Its hard and rectangular shaped and are available various varieties like milk , aniseed , whole wheat, condensed milk

Buns

Hot cross buns are a type of spicy, sweet yeast bread traditionally exchanged between Christians on Good Friday, particularly in England. They are also available at other times of the year, and are among the family of yeast leavened pastries available at most bakeries. Hot cross buns have a familiar and distinctive appearance, being puffy pastries quartered with a cross made from dough, icing, or simple slashes in the buns before they are baked. Traditionally, hot cross buns are baked together in a large pan and torn apart, creating a ragged edge along the sides of the bun.

Pastries

Pastry is the name given to various kinds of baked goods made from ingredients such as flour, butter, shortening, baking powder or eggs. It may also refer to the dough from which such baked goods are made. Pastry dough is rolled out thinly and used as a base for baked goods. Common pastry dishes include pies, tarts and quiches. Pastry is distinguished from bread by having a higher fat content, which contributes to a flaky or crumbly texture. A good pastry is light and airy and fatty, but firm enough to support the weight of the filling. When making a short crust pastry, care must be taken to blend the fat and flour thoroughly before adding any liquid. This ensures that the flour granules are adequately coated with fat and less likely to develop gluten. On the other hand, over mixing results in long gluten strands those toughen the pastry. In other types of pastry, such as Danish pastry and croissants, the characteristic flaky texture is achieved by repeatedly rolling out dough similar to that for yeast bread, spreading it with butter, and folding it to produce many thin layers.

Pretzel

These are dough pieces looped into various shapes such as knots and are baked with coating of salt. Pretzel has cracker like flavor, a crisp, brittle texture with brown glossy surface. With low moisture content it has longer shelf life. Basic ingredients for Pretzels are flour, water, yeast, sugar and shortening.

Biscuits and Cookies

The word biscuit comes from the Latin 'bis coctus', which means twice-baked. It is thought that biscuits have been baked for thousands of years and were originally baked in a hot oven and then cooled in a cool oven, although this process would not be found in modern processing factories. Cookie is derived from a Dutch word, koekje, which means little cake. The low moisture content of biscuits means they have a longer shelf life than other bakery products and so have been used in epic journeys such as sea voyages of the 15th century. British and European tradition involved serving biscuits in a semiformal situation with tea or coffee in between main meals, especially in the afternoons. Small biscuits were preferred so that a range of appearances and flavours could be offered without a large intake of food.

Biscuit is a term used in New Zealand (and Australia, UK, and South Africa) to describe a baked product that has a cereal base, e.g. wheat, oat or barley, of at least 60% and a low moisture content of 1–5%, excluding any moisture from fillings or icings. It usually has a higher fat content than other baked products, a longer shelf life and a higher energy density. In America the term for this product is cookie, while biscuit refers to a leavened bread-like product that is similar to the UK scone. Biscuits are usually made from hard dough, semi hard dough and soft dough. The term cookie is mostly used in American English. In British English, the term biscuit refers to a small, baked, unleavened cake, which is typically crisp, flat, and sweet. In North America, the term biscuit refers to a small, savoury cake that is somewhat similar to scones.

In the US, biscuits are classified based on their method of processing, especially they way in which they are shaped, with four main categories:

1. **Sheeting or cutting (also called cutting machine dough):** This method is used for hard dough, where it is passed through a series of rollers to obtain the desired thickness. The biscuit shapes are cut out of the sheets using a die which may be plastic or metal. The dough needs to be strong and elastic so that the biscuits hold their shape when the scrap is removed from around the cut biscuits.
2. **Rotary moulding:** This method is used for short dough and requires a dough with a relatively stiff consistency that is not sticky. The dough is compressed into dies mounted on the surface of a roller, with excess dough scrapped off. The moulded dough piece maintains it shape as it is pushed out of the die onto the baking sheet.
3. **Wire cutting:** Short dough is extruded through a die and sliced with a tight wire at appropriate intervals. The pressure placed on the dough in the extruder and the thickness of the wire vary dependent on the dough properties.
4. **Depositing:** Soft dough is shaped by depositing due to its semi fluid consistency and lack of cohesiveness. The dough is extruded through a nozzle and dropped onto a baking sheet. To achieve uniformity in the size and shape of biscuits, the flow of dough is cut off at regular intervals.

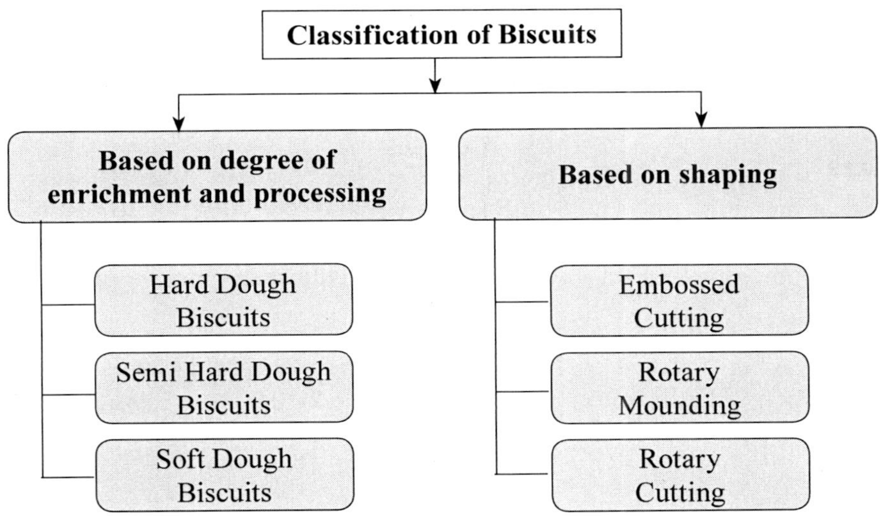

Fig. 1: Classification of Biscuits

Cakes

Cake is a leavened sweet baked food used as a dessert or snack made from flour, sugar, shortening, eggs, leavening agents and flavourings. Cakes are classified based on method of mixing and utilization of fat.

Classification of cakes

(a) **Shortened cakes:** Solid types of shortening agent like margarine and butter are used. Creaming and whipping are techniques are used for mixing the batter.

e.g., Velvet cake, Pound cake and Chocolate brownies

Brownies

Brownies is another popular bakery products often served with hot beverages like milk, tea or coffee. It's a flat, sliced mostly having chocolate as its main flavor hence the name brownie. Brownies are can be classified as cross between cake and cookie with variation such as fudgy or cakey. Normaly laced with ingredients like nuts, whipped cream, chocolate chips

(b) **Unshortened Cake/ Foam / Sponge type:** These are cakes made without addition of fat / shortening. Two basic types of unshortened cakes are Yellow sponge and White angel cakes. Yellow sponge is made with whole eggs whipping white and yellow separately. White Angel cakes are made with egg whites only. Beating, cutting and folding are used for mixing batter.

(c) **Chiffon type:** It is a combination of shortened and unshortened types. It uses liquid shortening like vegetable oil except coconut oil. Egg whites are separated and beaten. Cutting and folding techniques are used for mixing batter. Other popular bakery products include

Pizza

Pizza is a popular dish made with an oven-baked, flat, generally round bread that is covered with tomatoes or a tomato-based sauce and mozzarella cheese. Other toppings are added according to region, culture or personal preference.

Tortillas

Tortillas are unleavened bread made up of corn or wheat flour with water and salt popular in Latin America but used as staple food by Mexicans equivalent to Indian Naan ,Chapati ,Kulchas or Kaboos from Middle East.

Pies

Pies are baked dish containing pastry dough layer having sweet or sour fillings . Choco pies have recently become popular .Two crust covering filling s in complete shell.

Dumplings

Cooked balls of dough. They are mainly based on flour and could be made by steaming, frying, baking or boiling. Dumplings are stuffed with vegetables, meat or sweets. Dumplings can be sweet or savoury and served with stews or gravy or could be served by themselves

Types of Chocolate and Confectionery Products

Chocolate

Chocolate is derived from cocoa beans, which grow on cocoa trees in Central and South America, as well as in parts of the Caribbean and Africa. Cocoa butter is extracted from cocoa beans, as are cocoa solids. Cocoa powder is the name for cocoa solids that have been ground down into a powder. Cocoa butter contains all the fat of the cocoa bean, and it is used as the oil or fat with which chocolate is produced. Chocolate is made by mixing cocoa butter, which provides the fatty part of chocolate, with cocoa solids and sugar. The first step in producing chocolate is the fermentation of cocoa beans picked from cocoa trees. Cocoa beans have a naturally bitter taste when they are first picked, and must be processed a certain way to yield traditional, sweeter chocolate. The beans are then roasted, shelled, and ground into cocoa mass, which is often called chocolate liquor. Chocolate liquor can be broken down into cocoa solids and cocoa butter. The mixture of sugar, cocoa butter, and cocoa solids results in the kind of chocolate that we can buy at the grocery store. Many chocolate products are made by adding particular fruits, flavors, or nuts to traditional chocolate mixtures. These kinds of chocolate additives include almonds, cherries, hazelnuts, nougat, caramel, and marshmallow.

Chocolate defined as a food item that is rich in carbohydrates and sugar. Many governments, particularly those in European countries, enforce strict regulations in regards to the percentages of cocoa solids, cocoa butter, milk, or sugar used in chocolate production. Chocolate products can only be named and sold as milk, dark, or other types of chocolate if they meet these specific regulations for the percent of ingredients they contain. There is even an organization named the Chocolate Manufacturers Association, which deals with these kinds of regulations.

Many countries are famous for producing high-quality chocolate. These countries include Denmark, Italy, the United Kingdom, Germany, France, Spain, Mexico, and, perhaps most famously of all, Belgium and Switzerland.

Through the means of flavoring and processing, chocolate producers and manufacturers can create a huge variety of chocolate, all with different combinations of ingredients and levels of sweetness. The following are some of those varieties.

Dark Chocolate

Dark chocolate is identified as such because it contains far less milk than other kinds of chocolate, and sometimes no milk at all. It is made by mixing cocoa solids, fat, and sugar. Usually, chocolate with a cocoa solid component of 35% or more is deemed dark chocolate, but any percentage higher than that would cause the chocolate to be classified differently. Because of the absence of milk, dark chocolate appears far browner in color, and is sometimes referred to as black chocolate. Varieties of dark chocolate are often used in baking and cooking, because those recipes often include the addition of sugar, which is balanced out by the less intensely sweet flavor of dark chocolate. Dark chocolate is available in a range of cocoa percentages, and is sometimes sold with cocoa solids percentages of up to 90%. Dark chocolate is far less sweet than other varieties, including milk and white chocolate. Dark chocolate with higher percentages of cocoa solids will taste more bitter than those with lower percentages.

Milk Chocolate

Milk chocolate is probably the variety with which most people are familiar; milk chocolate is arguably the most popular kind of chocolate in terms of commercial consumption. It is made by adding milk, most often milk powder, to the traditional chocolate combination of cocoa solids, cocoa butter, sugar, and frequently vanilla flavoring. Before milk chocolate was available in solid form, people could only consume it as a liquid beverage. The use of condensed milk or milk powder, which is made by evaporating milk until only a powdered substance remains, enabled chocolatiers to create milk chocolate in solid bars, since powdered milk has a far longer shelf life than its liquid counterpart.

White Chocolate

White chocolate is a very sweet chocolate variety. White chocolate is unique because it is not made using any cocoa solids, and is therefore classified as a chocolate derivative rather than a traditional type of chocolate. White chocolate is a mixture of sugar, milk, and cocoa butter. The absence of cocoa solids and the presence of milk explain why this particular chocolate product appears ivory or yellow rather than brown in color. Interestingly, white chocolate will stay solid at a far higher temperature than milk or dark chocolate will, because the melting point of cocoa butter, its main ingredient, is very high. White chocolate lacks many of the antioxidant properties found in traditional chocolate, because it lacks the cocoa solid ingredients.

Unsweetened Chocolate

Unsweetened chocolate is a term often used interchangeably with baking chocolate. Unsweetened chocolate is produced without the addition of sugar, and thus it maintains more of its original flavor, which can be very rich and rather bitter. Chocolate liquor and fat are the two ingredients used to make unsweetened chocolate. Adding sugar or milk product would turn unsweetened chocolate into milk chocolate. Unsweetened chocolate is sold primarily for use in baking recipes, since its bitter flavor is balanced out by the large amount of sugar often called for in such recipes.

Semisweet Chocolate

Semisweet chocolate is technically a kind of dark chocolate. For dark chocolate to qualify as semisweet chocolate, it must contain half as much sugar as it does cocoa solids. Chocolate with any sugar to cocoa solid ratio larger than this one will be classified as sweetened chocolate. Similarly, bittersweet chocolate must contain sugar that amounts to no more than a third of the amount of chocolate liquor in the product. Like unsweetened chocolate, semisweet and bittersweet chocolate are most frequently used for baking and cooking purposes, and are commonly sold in larger quantities than other kinds of chocolate, such as blocks.

Compound Chocolate

Compound chocolate is made by combining cocoa solids with one of a number of substitutes for cocoa butter, including vegetable oil, coconut oil, and a variety of other hydrogenated fats. Many countries' chocolate regulations do not permit compound chocolate to be sold as traditional chocolate. Compound chocolate is often used as a topping or a coating for other Confectionery goods. Compound chocolate tends to be cheaper to produce and to purchase, because the substitution of oils or fats is less expensive than the process of adding cocoa butter.

Raw Chocolate

The term raw chocolate refers to chocolate that has not been processed in any way, or mixed with different ingredients. Raw chocolate contains only cocoa butter and cocoa powder, a form of cocoa solid. These ingredients are usually melted and combined to form raw chocolate. Raw chocolate is often sold as a healthy alternative to traditional chocolates, since it does not contain sugar or milk products.

Couverture Chocolate

Couverture is a relatively new term for chocolate that contains a high percentage of cocoa butter. This kind of chocolate is popular among gourmet chocolatiers and pastry chefs. Some semisweet and bittersweet chocolate varieties qualify as couverture chocolate.

Cocoa Powder

Cocoa powder is used mainly for baking and as chocolate drink mix, to which hot water or milk is added, to create hot cocoa or chocolate milk. Cocoa powder is a byproduct of chocolate production, and is sometimes used in its natural, unsweetened form for these purposes. Another kind of cocoa powder is called Dutch-process cocoa, which is made by removing the cocoa butter from pulverized chocolate liquor, and adding alkali to neutralize the acidic nature of the cocoa.

Sugar Confectinary products

Sugar confectionery refers to a large range of food items, commonly known as sweets. Boiled sweets, toffees, marshmallows, and fondant are all examples.

Sweets are a non-essential commodity, but are consumed by people from most income groups. The variety of products is enormous, ranging from cheap, individually-wrapped sweets, to those presented in boxes with sophisticated packaging.The main ingredient used in the production of sweets is sugar (sucrose). There is a danger that if sweets are consumed in excess over a prolonged period of time they may contribute to obesity. Unless good dental care is practiced, over-consumption can also lead to tooth decay.

By varying the ingredients used, the temperature of boiling, and the method of shaping, it is possible to make a wide variety of products. In all cases, however, the principles of production remains the same and include balancing the recipe, preparation of the ingredients, mixing of the ingredients, boiling the mixture until the desired temperature has been reached, cooling, shaping and packing. The degree of sucrose inversion, the time and temperature of boiling, the residual moisture content in the confectionery and the addition of other ingredients affect the storage of Confectionery products.

Sweets containing high concentrations of sugar (sucrose) may crystallize either during manufacture or on storage (commonly referred to as graining). Although this may be desirable for certain products (such as fondant and fudge), in most other cases it is seen as a quality defect.

When a sugar solution is heated, a certain percentage of sucrose breaks down to form 'invert sugar'. This invert sugar inhibits sucrose crystallization and increases the overall concentration of sugars in the mixture. This natural process of inversion, however, makes it difficult to accurately assess the degree of invert sugar that will be produced.

As a way of controlling the amount of inversion, certain ingredients, such as cream of tartar or citric acid, may be used. Such ingredients accelerate the breakdown of sucrose into invert sugar, and thereby increase the overall percentage of invert sugar in the solution. A more accurate method of ensuring the correct balance of invert sugar is to add glucose syrup, as this will directly increase the proportion of invert sugar in the mixture.

The amount of invert sugar in the sweet must be controlled, as too much may make the sweet prone to take up water from the air and become sticky. Very little will be insufficient to prevent crystallization of the sucrose. About 10-15 per cent of invert sugar is the amount required to give a non-crystalline product.

The temperature of boiling is very important, as it directly affects the final sugar concentration and moisture content of the sweet. For a fixed concentration of sugar, a mixture will boil at the same temperature at the same altitude above sea-level, and therefore each type of sweet has a different heating temperature (see chart below).

Table 2: Sugar Confectionery products and their temperature of boiling

Type of sweet	Temperature range for boiling (Degrees °C)
Fondants	116–121
Fudge	116
Caramels and regular toffee	118–132
Hard toffee (e.g. butterscotch)	146–154
Hard-boiled sweets	149–166

Variations in boiling temperature can make a difference between a sticky, cloudy sweet or a dry, clear sweet. An accurate way of measuring the temperature is to use a sugar thermometer. Other tests can be used to assess the temperature (for example, toffee temperatures can be estimated by removing a sample, cooling it in water, and examining it when cold). The temperatures are known by distinctive names such as 'soft ball', 'hard ball' etc., all of which refer to the consistency of the cold toffee.

The water left in the sweet will influence its storage behaviour and determine whether the product will dry out, or pick up, moisture. For sweets which contain more than 4 per cent moisture, it is likely that sucrose will crystallize on storage. The surface of the sweet will absorb water, the sucrose solution will subsequently weaken, and crystallization will occur at the surface - later spreading throughout the sweet.

The addition of certain ingredients can affect the temperature of boiling. For example, if liquid milk is used in the production of toffees, the moisture content of the mixture immediately increases, and will therefore require a longer boiling time in order to reach the desired moisture content. Added ingredients also have an effect on the shelf-life of the sweet. Toffees, caramels, and fudges, which contain milk-solids and fat, have a higher viscosity, which controls crystallization. On the other hand, the use of fats may make the sweet prone to rancidity, and consequently the shelf-life will be shortened.

Types of sugar confectionery products

(a) Fondants and creams

Fondant is made by boiling a sugar solution with the optional addition of glucose syrup. The mixture is boiled to a temperature in the range of 116–121°C, cooled, and then beaten in order to control the crystallization process and reduce the size of the crystals.

Creams are fondants which have been diluted with a weak sugar solution or water. These products are not very stable due to their high water content, and therefore have a shorter shelf-life than many other sugar confectionery products. Both fondants and creams are commonly used as soft centres for chocolates and other sweets.

(b) Gelatin sweets

These sweets include gums, jellies, pastilles, and marshmallows. They are distinct from other sweets as they have a rather spongy texture which is set by gelatin.

(c) Toffee and caramels

These are made from sugar solutions with the addition of ingredients such as milk-solids and fats. Toffees have lower moisture content than caramels and consequently have a harder texture. As the product does not need to be clear, it is possible to use unrefined sugar such as jaggery or gur, instead of white granular sugar.

(d) Hard-boiled sweets

These are made from a concentrated solution of sugar which has been heated and then cooled to form a solid mass containing less than 2 per cent moisture. Within this group of products there is a wide scope to create many different colours, flavours and shapes through the use of added flavourings and colourings.

References

Eliasson, A.C. and Larsson, K. (1993). Cereals in Breadmaking, Marcel Dekker, Inc. New York

Encyclopedia of Food Science and Technology, ed. F. J. Francis, volumes 1–4, John Wiley & Son, 1999.

FAO Repository, Cakes and Biscuits, FAO, United States of America

3

Bakery Ingredients

Baking is no different from any other area of cooking, and as in other sectors only the best and the freshest raw material can guarantee good results. So selection of right kind of ingredients is of utmost importance. Another basic need of a professional baker and confectioner is to purchase the equipment required. The design and size depends upon the volume of sale expected. So in this chapter we are going to learn about the ingredients required for running a bakery.

Table 1 : List of ingredients commonly used in bakery and confectionery

S. No.	Ingredients	
1.	Flour	(a) Wholemeal or whole wheat flour (b) Brown flour (c) White flour (d) Self raising flour (e) Strong flour (f) Soft flour
2.	Yeast	(a) Fresh (b) Dry
3.	Chemical Raising Agents	(a) Baking powder (b) Ammonium bicarbonate (c) Baking Soda
4.	Salt	—
5.	Corn flour	—
6.	Milk	(a) Liquid milk – full fat – low fat – skimmed (b) Milk powder (c) Condensed milk
7.	Cream	(a) full fat (b) low fat

(contd.)

S. No.	Ingredients	
8.	Shortening agents	(a) Butter (b) Margarine (c) Hydrogenated faWanaspati (d) Refined oil
9.	Egg	(a)Whole eggs (b) Egg whites (c) Egg yolks
10	Sugar	(a) Grain Sugar (b) Castor Sugar (c) Icing Sugar (d) Brown Sugar (e) Gold Syrup (f) Treacle (g) Liquid Glucose (h) Milk Sugar (i) Malt Sugar (j) Honey
11.	Coco and Covering chocolate	
12.	Coconut	
13.	Coffee	
14.	Nuts	(a) Almonds (b) Cashwenuts (c) Walnuts (d) Peanuts (e) Pistachio nuts
15.	Meat and poultry products	
16.	Fresh fruits and vegetables	
17.	Candied fruits	(a) Lemon and Orange Peel (b) Tutty Fruity
18.	Tinned/Canned fruits	(a) Cherries (c) Peaches (e) Banana (b) Pineapple (d) Mango (f) Fruit Cocktail (mixed)
19.	Spices and Aromatics	(a) Charmagaz (b) Cummin Seed (c) Nutmeg (d) Corriander Seeds (e) Red chilli powder (f) Cinnamom (g) Sesame seeds (h) Aniseed (i) Garlic

S. No.	Ingredients	
		(j) Cardamom - big and small (k) Poppy Seeds (l) Mace (m) Black pepper (n) Cloves (o) Onion seeds (p) Mixed spices (q) Ginger (r) Saffron
20	Alcohol	(a) Wines (b) Brandy (c) Rum (d) Liqueur
21.	Food Colours	
22.	Essences	(a) Vanilla (b) Orange (c) Lemon (d) Pineapple (e) Mango (f) Fruit Cocktail (mixed) (g) Cardamom - big and small (h) Poppy Seeds (i) Mace (j) Black pepper (k) Cloves (I) Onion seeds (m) Mixed spices (n) Ginger (o) Saffron (p) Rum (q) Liqueur (r) Strawberry (s) Pineapple

Selection, Storage and Use

The above ingredients can be classified into three categories depending upon their keeping quality, shelf life and the storage temperatures required.

(a) **Non-perishable -** Items that can be stored for more-than ‘a month at room temperature, e.g. flour, sugar, salt, spices, cocoa and coffee powders, colours and essences, canned products. They just require proper circulation of air in the storage area and protection from rodents and pests.

(b) **Perishable -** Items that can be stored for a couple of days at the most, at proper temperature, e.g. milk, cream, fresh fruits and vegetables, poultry and meat products.

(c) **Semi-perishable** - are those items which do not come under any of the first two categories, i.e. they require proper storage temperature. But the period of storage is more than that for perishables, e.g. butter and other fats, chocolates, tins/cans after opening, eggs. etc.

Flour

Flour is the most important ingredient without which production in a bakery or confectionery unit would be impossible. It is obtained by milling wheat. To understand flour properly it is important to learn about grain and its internal structure. During milling both bran and germ are removed. Bran has sharp edges which tend to cut the cell structure of loaf during proving, thereby affecting the volume of bread. Germ has more oil which affects the keeping quality of flour.

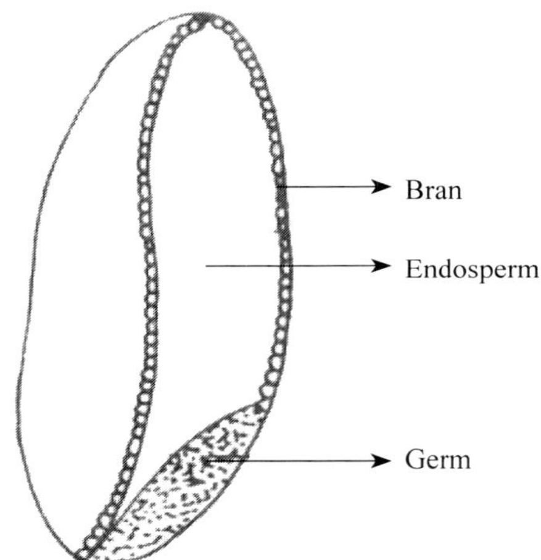

Figure : A wheat grain revealed

Table 2 : Different types of flour and its uses

S. No.	Whole Meal or Whole Wheat Flour	Brown Flour	White Flour	Self Raising	Strong Flour	Weak Flour
1.	Contains all parts of wheat grain, i.e., Bran, Endo-sperm	Coarser part of bran is removed	Most of the brain and germ is removed and contains mainly endosperm.	This flour has high-er protein content due to the wheat variety used or due to milling meth-od employed.	This flour has higher protein content due to the wheat variety milling used or due to method employed.	Contains higher proportion of starch and less protein.

S. No.	Whole Meal or Whole Wheat Flour	Brown Flour	White Flour	Self Raising	Strong Flour	Weak Flour
2.	Wholemeal flour is 'Atta' used for making chapatis and can be bought from provision stores.	Can be purchased from speciality food stores.	It is 'Maida' and can be bought from provision stores.	Available in speciality food stores.	Available in speciality food stores.	Available in speciality food stores.
3.	Used for making brown bread.	Used for brown bread.	Used for breads, buns, cakes, biscuits, depending upon their protein content.	Used for cakes and pastries.	Used for breads, buns, patties.	Suitable for cakes and biscuits.

Types of flour Different types of flour are used for different types of end products. Flours are identified as First Patent, Short Patent, Medium Patent and long Patent. Characteristics of these flours are determined by percentage of separation obtained from a 72% extraction. First Patent constitutes 70% separation from 72% extraction. Short Patent constitutes 80%, Medium Patent 90% and Long Patent 95% separation from 72% extraction. First Patent is used as cake flour and is obtained from soft wheats. Short Patent is used for premium brands of breads Medium Patent is used for featured brands of breads. Long Patent is used for competitive brands of breads

Flour Enzymes, Characteristics and pH value

Enzymes

Flour contains two enzymes which are essential to bread production. These enzymes are Beta-amylase and Alpha amylase. These enzymes develop in the wheat berry during the initial stages of sprouting. Beta-amylase converts dextrins and a portion of soluble starch to maltose which is essential for active yeast fermentation.

Characteristics

It is wish of every baker to use flour having the following characteristics for the production of quality bread: 1. Colour 2. Strength 3. Tolerance 4. High absorption 5. Uniformity

Colour: Flour should have a trace of creamish colour; otherwise, the bread will have a dead white crumb. Bleaching of flour contributes towards the control of degree of creaminess. In the case of bread it can be controlled by modification in baking formula,mechanical tratment of the dough, and by addition of ingredients which will effect the colour of flour.

Strength: It is often said that the flour is strong or the flour is weak. Such statement was refer to strength of flour which is capable of producing a bold, large volumed, well risen loaf. For the production of quality bread, strong flours need a longer fermentation period than weak flours. The quality of flour is decided by the amount of fermentation it will stand. Bread flour should be such that the dough made from it will retain its shape after being moulded.

Tolerance: The ability of flour to withstand the fermentation process and to produce a satisfactory loaf over a period of time, in excess of what normally is required to bring about the correct degree of ripeness for that practical process means tolerance.

High absorption: This refers to the ability of flour to take on and hold the maximum amount of moisture without additional mixing for full development of dough. It the dough is not given the required mixing time because of limited mixing capacity, or for other reasons, the baked product will lack volume and have a dry crumb and inferior eating and keeping qualities.

Uniformity: This is also an important aspect which should not be overlooked. If the flour lacks uniformity it will require constant checks and modification which are mostly considered as unproductive chores.

Yeast

Yeast is a unicellular micro-organism of plant origin. The biological name is *saccharomyces cervisiae.* Under right conditions the yeast increases by division and it is this process which makes yeast useful for baking. It needs air, moisture, warmth and nourishment (in the form of sugar) to multiply and produce carbondioxide to raise the dough. Yeast is available both in the fresh as well as the dried form.

Table 3: Differences between Fresh yeast and dried yeast

S. No.	Fresh Yeast	Dried Yeast
1.	Available from bakers and speciality food stores	Available from food and provision stores.
2.	Sold in 500 grams packs wrapped in butter paper.	Sold in small sachets.
3.	It is pale beige in colour, firm but crumbly in texture and ha a pleasant, fruity smell.	Available in granule form, colour is little darker than fresh yeast.

S. No.	Fresh Yeast	Dried Yeast
4.	Should be stored in a refrigerator loosely wrapped in a cling firm, can be kept for a couple of weeks. But it can be frozen for several months.	Until the packet is sealed it can be stored in any well ventilated, cool cupboard for a few months. But after opening the packet this yeast should be used within a month.
5.	Give better products	Quantity used should be half as compared to fresh yeast as this is more concentrated.

Yeast contains small amount of several enzymes including Protease, Lipase, Invertase, Maltase and Zymase. The important enzymes in yeast are Invertase, Maltase and Zymase.

Protease: The protease can soften flour proteins and therefore, it can cause great changes in the structure and properties of the dough. However, the Protease in normal healthy bakers yeast is intracellular enzyme.

Lipase: It seems to be intracellular and acts on the fats which it encounters within the yeast, especially during sporulation. These fats are the reserve products for the cell during the maturation of the reproductive spores. The Lipase of a few species of yeast is able to diffuse through the cell membrane.

Invertase: In most species of yeast, Invertase is an intracellular enzyme. The sucrose, cane sugar, enters the cell wall, is changed by Invertase to glucose and fructose which are simple sugars. These sugars then diffuse through the membrane.

Maltase: This enzyme, found in yeast, splits maltose sugar into two units of dextrose.

Zymase: Zymase is the enzyme that finally turns the trick in dough fermentation of sugars by yeast. It is encompassed by the complex group of enzymes collectively termed as Zymase. In bread production certain molds and bacteria can produce alcohols, but yeast is the most efficient agent.

Functions in dough making: The carbon dioxide produced by fermentation makes the dough rise. The carbon dioxide is a colourless, tasteless, edible gas obtained during fermentation or from a combination of soda and acid. In order for yeast to work at its best, the following requirements are needed:

(a) a balanced nutritional diet of sugar, nitrogen, minerals, vitamins and water and

(b) air, optimum environment of temperature, enzymes, oxygen, acidity, nutrient concentration and time.

Yeast derives fermentable sugar from four sources: (a) the yeast cell, (b) the flour, (c) hydrolysis of starch and (d) sugar added intentionally in the formula. Yeast ferments and grows best in acidic environment, tolerating as low as pH2 and then adjusting to its preferred pH near 4.5.

Chemical Raising Agents

(a) Baking Powder - is a mixture of sodium bi-carbonate, cream of tartar (tartric acid) and a separator, usually rice or potato or comstarch. Under the combined effect of air, moisture and warmth,carbon dioxide is produced from sodium bicarbonate whichagain causes fermentation. The separator prevents the two other ingredients from working prematurely by working as aninsulator. The acid present neutralizes the left over soda so that no after taste is left in the product.

Baking powder is the leavening agent produced by mixing an acid reacting material and sodium bicarbonate with or without starch or flour as filler. Since all baking powder must consist of baking soda the only way in which these can differ is in the type of acid ingredient used. Generally tartrate, Phosphate and Sulphate powders are used as acid ingredient. A baking powder should release its gas in the batter to saturate it with carbon dioxide gas and then liberate the gas uniformly during baking to hold the raised batter until set. This tends to give a uniform crumb and prevent shrinkage and cakes from falling.

Double acting Baking Powder: This type of powder consists of two acid ingredients- one fast acting and one slow acting. The fast acting part of this type of powder usually consists of calcium acid phosphate. These are the powders which release a good share of their leavening gas in a relatively short time after mixing and continue to release gas rather rapidly during the depositing operation and while the batter is on the bench. The slow acting ingredient will be either sodium pyrophosphate or sodium aluminium sulphate. These are the powders which do not release the greatest percentage of their leavening gas until the batter is heated in the oven. It is important to use a powder which has delayed action that a minimum of the gas is lost in the batter stage. This will be appreciated if one considers that a proper amount of carbon dioxide gas set free in the batter stage makes the batter smooth flowing and acids in scaling and panning.

Amount of Baking powder to be used: Amount of baking powder to be used depends upon the type of product, the character and amount of ingredients (like shortening, eggs) employed and the altitude of the place. It is highly important to be sure that the exact amount of baking powder required is carefully scaled. If the quantity of baking powder is excessive the cake will collapse or shrink after rising in the oven. The crumb colour of the cake will be dark and the taste will be foreign or salty. If too little is used the cake may not be sufficiently leavened resulting in a dense, heavy structure which lacks volume and good eating qualities.

Use of starch in Baking Powder: Filler – usually starch We have discussed the soda and acid part of baking powder so at this point we will discuss the filler. A filler in baking powder is necessary for several reasons. The most important of which are the following:

1. To keep the soda particles and acid particles from intimate contact and thereby minimizing the possibility of premature action.
2. To act as an absorptive medium for free moisture incorporated into the powder during manufacture or assimilated in the course of storage.
3. To facilitate handling and measuring in actual use. Pure white, redried corn starch meets all the requirements of a good filler and is the one usually used. Always use the freshest possible baking powder and store in a clean dry room with lid tightly covered when not in use.

(b) **Ammonium Carbonate and Ammonium Bicarbonate:** Either ammonium carbonate or ammonium bicarbonate is used t a small extent as a leavening agent. Its use is primarily limited to certain types of cookies. The advantage of this type of leavening agent is that it decomposes into two gases and does not leave a solid residue. There is distinct disadvantage in using either of above leavening agents in baked products with a higher moisture contents because of ammonia gas being soluble in water remains in the baked products and imparts to it a very disagreeable taste and odour. It is for this reason that the use of ammonia salts as leavening agents in baking is limited to those products which are small and porous units so that the ammonia fumes can bake out.

(c) **Baking Soda**: This is used frequently in commercial baking as it costs less. It contains sodium-bicarbonate which breaks into sodium carbonate, carbon dioxide and water. However residue of sodium carbonate leaves a bad taste and a dark colour which makes it not very suitable for most products except darker coloured chocolate cakes.

Salt

Chemical name of salt is sodium chloride. It contains 40% sodium and 60% chlorine. It is readily available in almost all parts of the world and is indispensable to cooking. Used by the bakers, it confers flavour and also accentuates other flavours. It has a stabilizing effect on gluten and controls the speed of fermentation in yeast aerated goods. It also helps on retaining moisture.

Cornflour

Chiefly produced from maize. It is white in colour and mainly contains starch which gelatinizes by mixing with water at a temperature above 170°F. Thus it is used as a thickening agent in custards and other confectionery items. It can also be used to dilute the strength of flour by mixing in suitable proportions.

Milk

It is a moistening agent and contains about 87% water. It is also an enriching agent depending on the amount used or whether it full fat, low fat or skimmed. Dried milk powder is very popular in baking because it occupies less space, keeps well if correctly stored, can be easily reconstituted or.can be sieved with flour and used in dry form. Condensed milk is produced by evaporation of water under vacuum. It is generally sweetened.

Functions of Milk

While using milk in bakery products, it should be considered in two parts. These refer to 1) water 2) total solid contents in milk. The water in liquid milk could range between 12.5% to 90% depending on the type of milk. The water has a number of functions when present in proper quantities. It contributes towards eating qualities. The water in milk combines with the other ingredients. In order to have flour develop structure water is absolutely necessary. In order for sugar to be a tenderizer, moisture must be present. Thus the moisture of the milk is neither a toughener not a tenderizer, but when combined with other ingredients may contribute to both toughness and tenderness in the products. The milk solids have a binding effect on the flour protein, creating a toughening effect. They also contain lactose which helps to regulate crust color. They improve the flavour and are important moisture retaining agents. If liquid forms of milk are used, the water content should be taken into consideration for adjusting the formula.

Advantages of using Milk solids in Bread Production: There are several advantages that could be derived from adding milk solids in the bread dough.

1. Increased Absorption and Dough Strengthening
2. Increased Mixing Tolerance
3. Longer Fermentation
4. Better Crust Colour
5. Better Grain and Texture
6. Increased Loaf Volume

7. Better keeping Quality
8. Better Nutrition

Cream

It is used in cakes, desserts and for decoration. It is the skimmed milk fat and has a pleasant flavour. The creams vary in thickness and richness. The higher the butter-fat content the less likely cream is to fall after whipping. Fresh milk and cream both need to be stored in the refrigerator and cannot be stored for more than a few days.

Table 4 : Different types of shortening agents

S. No.	Butter	Margarine	Vanaspati Ghee	Refined Oil
1.	Made by chuming milk fat.	It is cheaper butter substitute made from hydrogenated oils.	Made by hydrogenenating vegetable oils.	Consist of 100% fats with low melting point.
2.	Contains about 85% fat and rest is water and milk protein	Controls hydrogenatd oils, ripened milk, colour and salt.	Contains mainly fats.	A liquid room temperature it cannot be used for creaming.
3.	Has a pleasant aroma and good for bakery due to this.	Has no aroma but rest of the physical characteristics similar to butter.	Has grainy texture and no aroma. Due to the grainy tex-ture less suitable for baked goods.	Mainly used for frying and tin greasing.
4.	Butter should be firm and it should not be stored at a temperature below 40°F.	Can be stored at room temperature.	Can be stored at room temperature.	Can be stored at room temperature.

Eggs

After flour, eggs are the second structure forming materials used by the baker. Both, egg white and yellow are of great importance. Egg white whisks easily and makes cakes and pudding lighter. During baking it solidifies to lock in the air. Egg yolk emulsifies well and is used as a glaze and also in ice-creams and cream desserts. An average egg weighs around 45–50 g. A fresh egg sinks in water whereas a stale one floats. The yolk of the egg should be firm. Egg can be stored in the refrigerator for a week or two.

Sugar

(a) **Grain Sugar :** This is the sugar we use normally at home. It contains 99% water soluble carbohydrates and 1% water.

(b) **Castor Sugar :** is a finer form of granulated sugar and is suitable for creaming in baking.

(c) **Icing Sugar:** It is a very finely powdered white sugar which is used for icing, glazes, dusting cakes after baking and for almond paste.

(d) **Brown Sugar:** These are the un-refined raw sugars, some having names that refer to country of origin, e.g. Barbados, Demerara, etc. All brown sugars confer colour and some flavour. These sugars are ideal for rich cakes.

(e) **Golden Syrup:** This amber coloured syrup is a by-product of sugar refining. It is used by the baker for ginger cakes and biscuits.

(f) **Honey:** It is a thick natural syrup obtained by bees from the nectar of flowers. It is used in fresh ginger breads, nuggets etc.

(g) **Treacle:** It is a syrup much darker in colour and with a more pronounced flavour than golden syrup. It is made by diluting and filtering molasses and then concentrating. Treacle can be used for ginger goods, dark heavily fruited cakes and Christmas pudding. The treacle replaces some of the sugar in the mixture.

(h) **Liquid Glucose/Corn syrup:** It is made by boiling starch in water so that it is gelatinized. A weak acid is added to the gel to get sugar. It is used in cakes and biscuits and in sugar boiling.

(i) **Milk Sugar:** Milk sugar or lactose is obtained from fresh and skimmed milk. It is used to impart additional flavour and sweetness.

(j) **Malt Sugar:** Malt sugar or maltose is obtained from milk syrup and adds sweetness.

Cocoa and Chocolate

Both are obtained from cocoa beans. Cocoa powder is low fat and has no sugar whereas chocolate has some sugar, cocoa, butter and milk added in varying quantities. Both cocoa powder and chocolate are used considerably in confectionery products. Cocoa powder can be stored in air tight containers in well ventilated places for months. Chocolate should be wrapped in polythene paper or aluminium foil and then refrigerated.

Cocoa and chocolate are used very widely in the production and finishing of cakes, pastries, pies and cookies. They provide for variety of products and the characteristic flavour and colour in the product and also supply body and bulk to the cake mix or icing. Chocolate and Cocoa are produced from cocoa beans which are the fruit or seed of the cacoa trees. Cacao trees grow in tropical areas mainly- Ghana, Indonesia, Brazil, Sri Lanka and Venezuela. The cacao bean is the source of cocoa chocolate. The cacao pod is cut from the tree and split open. The beans inside are picked, washed, dried and fibres removed before they are fermented and cured. This process gives the bean the aroma, taste, flavour and coloour. The beans are then shipped to the processor where they are roasted for further improvement in the taste, flavour and colour. After cooling, the beans pass through rollers to produce cocoa nibs. This action loosens the husks which are discarded. The nibs are crushed and reduced to a thick mass called chocolate liquor. The chocolate liquor contains approximately 54% cocoa butter. Cocoa butter is removed by pressing melted chocolate liquor in filter presses to obtain the cocoa butter needed in the manufacture of sweet chocolate coatings. The dry looking residue in the press is reground to make cocoa powder.

Dutch-Process cocoa

Dutch-process cocoa is a modified form of natural cocoa. In this process the cocoa beans are treated with an alkali. A solution of alkali (usually sodium bicarbonate) is added to the partially roasted beans and the excess water is evaporated off. The beans are then treated as for the manufacture of natural cocoa. The purpose of dutching the cocoa is to darken the colour, modify the flavour and to improve the dispersibility of the natural cocoa. This process also reduces the natural acidity of the cocoa because the alkali neutralizes some of the acidity. Dutching not only darkens the colour, but also makes the cocoa less bitter and easily dispensible.

Natural Cocoa

Natural Cocoa which is known as the breakfast cocoa has a lower pH (4-5) or greater acidity and requires the use of the more baking soda as the leavening in the mix than the dutched cocoa. The dutched cocoa does not have a strong or concentrated flavour. It has a high pH value (6-8.8) or excess alkalinity. This cocoa requires less baking soda for leavening and more baking powder. The type of cocoa used should be known in order to make adjustments in the leavening depending upon the nature of cake. If the cakes are mahogany or reddish brown it is a sure indication that too much soda quantity was used. The cake will also have a baking soda or soapy taste. Adjust the soda quantity used in the mix. If the cakes are light brown or cinnamon coloured then they require an increase in the amount of soda.

Coffee

Coffee is an excellent flavouring for creams, fillings and icings.

Nuts And Dried Fruits

These are of great importance in cakes, pastries and puddings. Walnut, pistachio nut,

A variety of dried and preserved fruits and nuts can be used in baked products to produce different types of flavours and finishes. It is usually necessary to wash dried fruits before use with a liberal amount of water and swirled around for about one minute. Care must be taken so that the fruit does not absorb too much water and become soft. If fruit absorbs too much water it will break down during mixing and discolour the dough. The flavour also diminishes if the fruit is soaked too long. After washing the fruits should be drained in sieve. After draining the fruits should be carefully picked over by spreading the fruit on a dry cloth to remove the excess moisture. The fruit should always be added last to ensure even distribution throughout the batter / dough with minimum damage.

Dried fruits: Among all the dried fruits, the products of different types of grape vines take the foremost place in the confectionery.

Currants

Currants are the dried form of small black grapes. Good quality currants should be bold, fleshy and clean of even size and blue black in colour. The currant should not contain red shriveled berries which due to their extra acidity spoil the flavour of cakes. Currants prior to their use should be soaked in boiling water for 2 minutes and then dried to get rid of the frit, stalks and stones.

Sultanas: Sultanas are made out of seedless yellow grapes.

Raisins: Ripe grapes are converted into raisins and sultanas in different ways. In the case of raisins, the bunches of grapes are partly dried by twisting the stalks while still on the vine and then are finished off in open sheds.

Dates

Dates are dried fruits of Iraq and North African palms. They are very sweet and rich in sugar. Dates should always be soaked in about half of their weight of water for an hour or more until they are soft. Sugar preserved Fruits and Peels

The skin or peel of the citrus fruits such as lemons, oranges and citrons are abundantly used in fruit cakes. Thick rind fruits are the best for the preparation of peels. These fruits are cut across the middle and the pulp is removed. The halves known as caps are soaked in brine for several days to remove the acid taste of the rind.

Cherries

Glace cherries

Good quality fruit is bleached in a solution of water, calcium carbonate and sulphur dioxide until the fruit is colourless. The fruit is stoned and washed. Cherries are then soaked for a few minutes to soften the skin and flesh. After draining, cherries are immersed in weak syrup which is coloured usually red, but also green or yellow. The syrup strength is increased daily by boiling over a period of 9 days.

Crystallized Cherries

The crystallized cherries are made by draining the preserved glaced cherries and rolling them in fine castor sugar. Crystallized fruits Crystallized fruits are mostly used in the decoration of rich fruit and other cakes. Before using, the crystalline sugar from the fruits is washed off and the fruits are cut to the desired shapes and placed on the cake. After baking the fruits can be washed with a good syrup to enhance the brightness of the fruit. Pineapple, peaches, apricots, plums and pears are generally used to make crystallized fruits.

Angelica

Angelica is large green plant of which only the stem is used. It is preserved in a similar way to cherries, using green syrup. Apart from the bright green colour it has an aromatic flavour.

Ginger Root

Only the tuberous root of the plant is used. It is washed well and boiled in a weak sugar solution until soft. The sugar strength of the syrup is to be gradually increased as in the candying process of citrus peels. It is stored in syrup as root, chips or crushed.

Crystallized Flowers

Rose petals are laid out on wires and suitably coloured syrup is allowed to drip on to them. When thoroughly saturated, the petals are dried over gentle heat. They can be used as a decoration, owing to the colours but also have a scented flavour of the original flower.

Nuts

Nuts offer various flavour, texture, bite and appearance in baked products especially in the products of cookies. There are several nuts available in various shapes and all have a high food value. Most of the nuts are expensive which restricts their use only in specialty items.

Almonds

Among the nuts the almonds have richness and fineness of flavour but due to their high cost are used in specialty items and for decorations. There are two types of almonds: the sweet and the bitter. The bitter almond is used in the preparation of essential oil of almond. Bitter almond is not suitable for eating and is used to boost up the flavour by blending with sweet almonds. In Confectionery almonds are used either whole or split or ground or a combination thereof according to the type of desired end products.

Walnuts

Walnuts have a strong flavour and due to high fat content have a tendency to go rancid after a long storage.

Pistachio Nuts

Pistachio Nuts are sparingly used in the baked products because of their high price. They are about ½ — in length and have a purplish brown skin which can be removed by blanching.

Cashew Nuts

They have a bland flavour. They are largely used for decorative purposes.

Groundnuts (Monkey Nuts or Peanuts) They contain about 40% of oil which is used in making vegetable fats. Because of price advantage, groundnuts are generally used in various cookie recipes.

Coco Nuts

The white flesh known as copra is removed from the shell of the nut and is dried either in the sun or in the shade. A better colour is produced by shade drying. When dried, the copra is cut according to confectioners requirements such as shredded, coarse, medium or fine dessicated coconut. Cut coconut can also be coloured by mixing well with a liquid colour and then drying off the excess moisture. Owing to high oil content, coconut is liable to develop rancidity. Coconut has a tendency to be contaminated with salmonella bacteria which is harmful for health.

Meat and Poultry Products

These are used as fillings for savoury items like patties, vol-au-vent, pizza, barquettes etc. They should always be fresh and of good quality. These need to be refrigerated, if stored for a couple of days.

Fresh Fruits and Vegetable

Fresh fruits and vegetables form an integral part of any bakery. They should be fresh when used.

Food Colours

The use of colour is important as the use of flavour. The eyes appeal of the product is enhanced by the use of colour. The correct colour should be used to complement directly the flavour added in the product. The following are the guidelines for the use of colouring materials: a) To supplement deficiencies in colour, e.g yellow colour is used to conceal the lack of butter and eggs in a dough. b) To increase the eye appeal and to complement a definite flavour. c) To introduce varieties and interest to decorated products.

Natural Colours

Cochineal or Carmine

Cochineal is a red from which many pinks are derived. It is prepared from an insect which lives on a variety of cactus originating in Mexico and cultivated in the Canary Isles. The dried insects are powdered and then boiled to extract the colour. The filtered liquid is known as Cochineal. Carmine is prepared from this liquid by the addition of acid.

Saffron

Saffron is a yellow to orange colour prepared from the flowers of a crocus grown in southern Europe and Kashmir. The dried stigmas are available in cake forms or loose. The colour and flavour of saffron is extracted by making a hot water infusion which should be prepared when required, since the colour soon decomposes.

Turmeric

Turmeric is prepared from the dried and ground root of the ginger family plant grown in India. Its colour closely resembles eggs.

Annatto

Annatto is yellow colour prepared from the fermented fruits of a plant grown in the West Indies and Sri Lanka. The colour is hardly soluble in water but is readily in alcohol. It is used for colouring butter, cheese and margarine.

Chlorophyll

The green colouring matter chlorophyll is extracted from the leaves of Spinach etc.

Caramel

Caramel or black jack is a dark brown colour prepared by heating sugar until it is decomposed and then adding boiling water to form a thick syrup. Caramel will impart a distinctive flavour of burnt sugar

Setting Materials

For the manufacture of jams, jellies and icing, ingredients are required either to cause thickening of a fruit syrup (jam or jelly) to set a firm gel or in preventing stickiness of icings. These ingredients are known as gums. Most of the gums are carbohydrates from sundry plant sources as sap, seeds and sea weeds and are known as plant gums. Gelatin is an animal protein which is not a gum but is included as a stabilizer.

Gelatin

Gelatin is derived from cartilage or bone of the animals. Pure gelatin is transparent brittle substance without colour, smell or taste. It is available in sheet form and as crystalline powder. It has a longer shelf life if it remains dry. When moistened it will deteriorate rapidly. It dissolves only in boiling water. When placed in hot water it swells and will absorb about ten times its own weight. Solution is thin bodied when hot, but at 1% concentration will set on cooling. Prolonged boiling destroys the gelling properties. Excessive amounts make products tough and rubbery.

Agar-Agar

Agar-Agar is a jellying agent derived from a variety of seaweed available in southern Asiatic waters. It is available in powdered or fibrous form. It is also called vegetable gelatin. It is insoluble in cold water but will absorb large quantities of water and when dissolved in boiling water produces a slightly cloudy gel which is slightly less tough than the one formed from gelatine. At 10% concentration the solution will set on cooling. Boiling is necessary to dissolve agar-agar but boiling does not detract from its setting powers.

Pectin

Pectin is the setting agent used in the production of jams. It is present natural to a greater or lesser degree in most fruits. It is water soluble which when in solution the presence of sugar and acid is capable of forming a gel. To obtain a firm setting it is necessary that the fruit contains not only. Pectin but also acid. Fruits that are deficient in either or both can be supplemented by the addition of acid or Pectin or by a second fruit that is rich in either or both. Commercially Pectin is prepared from citrus fruit pulp or from papaya, guava etc.,

Isinglass

Isinglass is an extremely pure form of gelatin obtained from the swimming bladder of various fish. It dissolves in hot water and sets as a jelly when cool. Due to its high price it is not widely used.

Irish Moss or Carrageenan

Irish moss is also derived from seaweed and is widely used as a setting or stabilizing agent in emulsions. It is generally used for setting the milk products because it reacts with the casein of the milk to form excellent firm but tender gel.

Water

Water, like any other common ingredient of bakery products, must be uniform in order to obtain uniform results in the products, If the water supply is constant in hardness and pH, then once the formula has been adjusted to meet the requirements of the water there probably will be no problems from the water. If, however, the water supply varies then adjustment in the formula will need to be made.

Types of Water

Water may be classified into six different headings consisting of soft water, hard water, Alkaline water, Acid water and Turbid water. Soft water has low content of dissolved minerals whereas Hard water contains dissolved minerals in an appreciable amount. Hard water can either be of temporary hardness or of permanent hardness. When alkali is present in the soil, water tends to become alkaline. Acid water is often found in areas where there are mines, mine wastes, and in water receiving the waste from industrial processes. Acid water is a rarity from a natural source. Saline water contains traces of common salt there by making it sensitive to taste.

Functions of Water in Baking

Water has functions in bread making. It makes possible the formation of gluten. Gluten as such does not exist in flour. Only when flour proteins are hydrated, gluten is formed. Water controls the consistency of dough. Water assists in the control of dough temperatures and warming or cooling of doughs can be regulated through water. It dissolve salts; suspends and distributes non-flour ingredients uniformly. Water wets and swells starch and renders it digestible. Water also makes possible enzyme activity. Water keeps bread palatable longer if sufficient water is allowed to remain in the finished loaf.

Flavourings

Flavour may be defined as the sensation of smell and taste mingled. Flavour is an important ingredient in a sweet goods formula. Flavour is really the ingredient which helps the baker to add uniqueness to his product. Appearance may be an eye-catching factor in the first sale of any baked product but flavour holds the key to all subsequent sales. The general accepted components of taste are: Sweetness, sourness, saltiness and bitterness. There are various sources through which the baked product can acquire its unique bakery flavour. It can acquire the flavour during the processing of the product, i. e during baking, fermentation etc.,

a. Fermentation

The total fermentation time has a profound influence upon the end flavour of the baked product due to the biological changes that take place during the fermentation. Breads made from sour dough or overnight sponge has a different flavour from the breads made from short sponges and straight dough process.

b. Baking

The process of baking brings about two important changes which add flavour to the product:

i. Brown reaction
ii. Caramelization

Flavour Additives

These additives can be divided into three groups: Natural, synthetic and imitation (with unlimited combination of all three)

1. **Natual:** a. Basic ingredients added to the formula: forms of sugar and syrups i.e honey, molasses malt syrup etc, ground fresh fruit, cocoa, chocolate etc., b. The essential oil of citrus fruits such as oil of lemon and oil of orange and vanilla extract
2. **Synthetic:** The quantities of flavours present in the fresh fruit are very small. If the flavour from the fresh fruit was to be used singly alone in the formula, large quantities of ground, sliced fruit will be necessary to bring about the desired level of flavour. This will not only unbalance the formula but will make it impracticable. If this natural flavour is fortified with synthetic flavour it will have more taste appeal than the use of natural flavours alone.
3. **Imitation**: The imitation of natural flavours are rarely used alone but are blended with fruit juices and essential oils to give a better result. Imitation flavours are not found in nature but used to reproduce the natural flavour.

Classification of Flavours:

i. **Non-alcoholic flavour:** These are prepared by dissolving ingredients in Glycerine propylene, glycol or vegetable oil. These help to retain the flavour during baking by reducing vaporization.

ii. **Alcoholic extract:** These flavours are dissolved in ethyl alcohol. They are too volatile for use during baking but are very suitable for icing and fillings which do not undergo baking.

iii. **Emulsifiers:** These flavouring oils are dispersed in gum solution which help to obtain an even distribution through batter and dough and also maintaining stability during baking.

iv. **Powdered flavourings**: These are prepared by emulsifying components in heavy gum/water solution, then spray dried to form powders.

References

Biscuit Cookie and Cracker Manufacturing Manual 1 Ingredients, Woodhead Publishing, Cambridge, 1998

Cereal Processing Technology, ed. G. Owens, Woodhead Publishing, Cambridge, 2001.

Gisslin, W. Professional baking. New York : John Wiley & Sons, c1985.

4
Formulations and Processing of Biscuits

The word biscuit is derived from *Danis biscoctus* which is Latin for twice cooked bread and refers to bread rusks that were made for mariners (ships biscuits) since the Middle Ages. The dough pieces were baked and then dried out in another cooler oven. All biscuits are made with flour (usually wheat flour) and all have low moisture content and thereby have long shelf life if protected from moisture and oxygen. It includes items also known as crackers (that make a noise of cracking when broken), Hard sweet and semi –sweet biscuits, cookies (which is the name that originated from Dutch word *koekje* meaning a small cake) and wafers, which are baked between hot plates from a fluid batter. The name cookie was adopted in North America where the term 'biscuit' can be confused with small soda raised breads or muffins. In other countries the term cookie is used primarily for wire cut products of rather rough shape, which often contain large pieces of various ingredients such as nuts, etc.

The term 'biscuit' is used in Britain to describe a flat, crisp, baked product; the term cookie is reserved for something softer and thicker. Cookies are made from soft wheat flour and are characterized by formula high in sugar and shortening and relatively low in water. Similar product is known as biscuit in our country. In USA the term 'cookie' covers any flat, crisp, baked good. Cracker is a term reserved for biscuit of low sugar and fat content, frequent bland or savoury. Crackers are usually made from developed dough whereas cookies are made from weaker flour. These foods have in common their ability to stay palatable for a long period of time. Their ease of transportation and their ready consumption without any further preparation make them high on the list of staple snacks

Ingredients Used in Biscuit Making

Flour

Flour constitutes the primary raw material to which all soft wheat product formulations are related. It provides a matrix around which other ingredients in varying proportions are mixed to form better or dough systems. Most biscuits can be prepared from flour, which has low quantity of protein and has a gluten content that is weak and extensible. Thus flour with protein level of less than 9% is best and levels of more than 9.5% often create processing problems. The exceptions are fermented cracker doughs and puffs doughs where a medium strength of flour is needed, with protein value of 10.5% or more.

Different types of flours are used ranging from soft to strong sponge flour in the3 bakery product processing. Cookie, cracker flours normally receive no special treatment or additives i.e. they are not normally chlorinated or chemically matured and have no chemical leavening additives or self-raising ingredients. High protein in the flour leads to hardness of texture and coarseness of internal grain and surface appearance. Flour should be sifted to aerate it for easy mixing operation.

Table 1: Typical specification of flour used for chemically aerated biscuits.

Wheat types	A maximum content of soft wheat varieties
Moisture content	14%
Smell	Free from mustiness (mould), no taints from paints, detergents
Protein (Nx5.7), %	9.0% + 0.5
Colour grade figure (Kent -Jones)	3.5 + 1.0
Ash content Particle size range	0.46 – 0.55%
Particles greater than	250 ppm < 1.1 %
Particles greater than 50 ppm	< 40%

Sweetener

It imparts sweet taste, improves texture, crust colour and extends shelf-life. Selection of the proper sweetener mostly is determined by the desired functions the sweetener is to provide. The principal sweetener used is sucrose (granulated sugar). Corn syrup, high fructose com syrup, invert sugar, honey, glucose syrups and molasses are used to a lesser extent except in soft cookies. Granulation of sugar is very important. Coarse grain of sugar will cause more spread of cookie affecting its texture, eating quality etc. Very fine granulation will not incorporate enough aeration resulting in dense texture, toughness and poor eating quality. Coarsely powdered or a fine granulated sugar should be used. Some cookies are moist in eating in which case part of sugar is replaced with liquid in which case part of sugar is replaced with liquid sugar such as invert sugar, honey or com syrup. Dextrose sugar will have reduced sweetness and will impart darker colour to the cookie.

Fat/shortening

Fat lubricates the structure of a baked product. It has tenderizing effect on flour proteins and makes the product tender. Fat improves the eating quality for prolonged period. If the fat level is high the lubricating function in the dough is so pronounced that de little or no water is required to achieve a desired consistency, little gluten is formed and of starch swelling and gelatinization is also reduced giving a very soft texture. The dough and breaks easily when pulled, it is short. This is the origin *of* the term 'shortening' for a dough fat. Generally smooth, plastic hydrogenated shortenings are used for cookie making.

Granular shortenings are unsuitable as they do not aerate sufficiently and distribution *of* fat in the cookie remains uneven. The fat used for cookie making should be able to cream and incorporate aeration and should not melt at baking temperature. Fats used as surface coatings applied as a spray *of* warm oil, for savory crackers are best if they have limited absorption into the cookie and remain as a glossy film. Addition *of* part butter improves the taste and flavor of the cookie and also does not affect the creaming quality Butter should be softened before blending with hydrogenated fat otherwise it will break into lumps which will be difficult to homogenize.

Eggs

Eggs affect the texture in several ways. They perform emulsifying, tenderizing and binding functions. Eggs also contribute colour, nutritional value, and desirable flavor. They are essential for obtaining characteristic organoleptic qualities *of* products. Eggs whites are a toughener and structure builder and the high fat contents *of* yolk function as a tenderizer. Eggs must be fresh. Stale eggs may give bad odour and spoil the overall flavour *of* cookie. Whole eggs are best used at room temperature while egg white whip better when it is cooled. Egg yolk alone or in combination with whole egg produces a cookie with excellent eating quality with a bit inferior grain or internal structure compared to that from whole egg.

Milk

Milk is generally used in the form of dry milk non-fat. It imparts good colour, flavor and a very creamy eating quality. One or two percent of milk solids achieve very desirable results. Dry milk is best used after dissolving in water if, water is an ingredient of the formula. Milk powder should be mixed with equal quantity of sugar in dry state and then small quantity of water should be added to make lump free slurry. It can also be sifted along with other dry ingredients.

Flavours

Choice of flavours in cookies is very limited. Generally use of butter and milk as the ingredients of the formula perform the function of flavoring agents, which is further fortified with vanilla, which is used within limits of 0.5 to 1 percent based on flour. Some spices like cinnamon, nutmeg, ginger, jeera are also used as flavours. Flavours should be used with utmost care as even slightly enhanced quantity may impart very strong and unacceptable flavor to the product.

The introduction of aromatic ingredients as a contribution to flavor can be made to biscuits and other cooked products in three principal ways:

1. By including the flavor in the dough or batter before baking
2. By dusting or spraying the flavor after baking
3. By flavoring a non -baked portion, such as cream filling, icing, jam or mallow, which is applied later.

Other ingredients

Ingredients that play a texture-modifying role include raising agents and emulsifiers. Raising agents lighten the structure of a baked product during, baking by releasing tiny bubbles of gas in the dough or batter, which expand as the temperature rises, opening up the products structure and thereby lightening the texture. Baking powder, bicarbonates of soda and ammonia are commonly used raising agents.

Baking powder

Baking powder is combination of sodium bicarbonate and an acid salt (phosphates, tartrates, sulphates) when moistened and heating will evolve gas, which leaven the product giving it volume and making light and easy to digest. Baking powder must yield not less than 12% available carbon dioxide. The reactivity of baking powder is determined by their neutralization value (NV), which is defined as the numbers of grams of soda that 100 g of acidic salt will neutralize. Baking powders are classified as 'fast acting' 'slow acting' and 'double acting'. Fast acting powders release most of their gas at room temperature. Slow acting powders release a portion of the available carbon dioxide during mixing but generate most of it by reactions occurring at elevated temperatures. Double acting powders are version of the slow acting type that has somewhat more gas producing potential during mixing. This type of baking powder is most widely used by bakers (Table 2).

Baking soda

It is chemically known as sodium bicarbonate. It will liberate CO_2 gas, a leavening gas, when heated. It also liberates the same gas when mixed with an acid, either hot or cold. The popularity of sodium bicarbonate as a gas source is based on its low cost, lack of toxicity, ease of handling very small contribution to the taste of the end product.

Ammonium bicarbonate

When ammonium carbonate or bicarbonate is heated CO_2 and NH_3 is produced. No solid is left behind in this reaction. However, the ammonia imparts a detectable odour unless it is completely removed. It is used in biscuits and crackers as they have large surface to mass ratio and ammonia escapes when baked at high temperature. It can be used in products that are to be baked at low moisture.

Table 2: Types of Baking Powder

Type	Ingredient	Percentage
Fast action:		
Formula No. 1	Tartaric acid	5.97
	Cream of tartar	44.90
	Sodium bicarbonate	26.93
	Starch	22.40
Formula No. 2	Monocalcium phosphate	33.43
	Sodium bicarbonate	26.73
	Starch	39.87
Slow action		
	Sodium acid pyrophosphate	40.38
	Sodium bicarbonate	30.59
	Starch	29.03
Double action		
Formula No. 1	Monocalcium phosphate	13.28
	Sodium aluminium sulphate	19.92
	Sodium bicarbonate	26.73
	Starch	40.07

Classification of Biscuits

Biscuits may be classified in various ways.

1. Based on the texture and hardness.
2. Based on the method of forming dough and dough pieces e.g. fermented, develop laminated, cut, moulded extruded, deposited, wire cut, co extruded etc
3. The enrichment of recipe based on fat and sugar.

Based on formulation biscuits have been classified based on the dough consistency such as hard dough biscuits and soft dough biscuits.

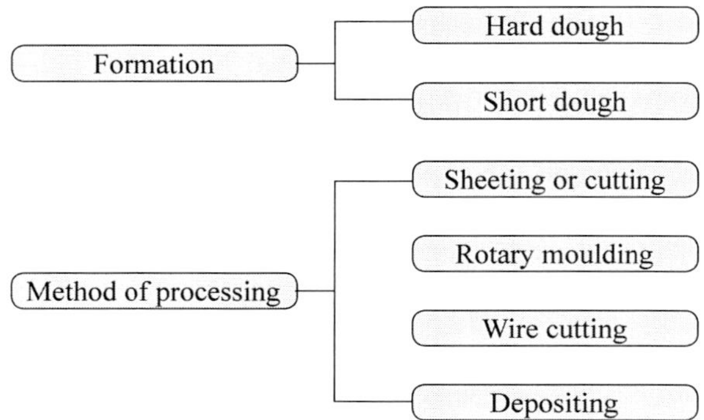

Source: Wrigley et al., Encyclopedia of Grain Science, Cookies, Biscuits, and Crackers, 2002.

Fig 4.1. Classification of Biscuits

Hard Dough Biscuits

In hard doughs the gluten is partially developed and to some extent extensible depending on the percentage of sugar and fat in the composition. In this category the biscuits that can be included are:

(i) Water biscuits

(ii) Sweet gluco biscuits

(iii) Semi sweet Marie type or cabin biscuits and some of the specialty biscuits having slightly higher percentage of shortening

(i) Water biscuits

Water biscuits have a simple recipe mostly of flour, fat, salt and water in the ratio of 100:6.5:1:29. The dough is under developed and crumbly or in balls after mixing. They may be then given the conditioning period before sheeting when some form of proteolytic activity mallows the gluten to make it more extensible. A dough sheet is formed which after laminating is cut and baked in a very hot oven Water biscuits are usually round and may be as large 3:S 70 mm in diameter. As longitude shrinkage occurs in the oven, the cutter must be oval and the shape is controlled by the relaxation of the dough before cutter. Mostly Jewish community prepares water biscuits. No flavor is added in biscuits.

Water biscuits Formula

Flour - 100; Fat - 6.5; Salt - 1.0; Water - 29

(ii) Gluco biscuits

In India gluco biscuits are manufactured in the largest quantities and because of lower cost it is most popular among children. A typical recipe of these biscuits is given below.

Gluco biscuits formula

Wheat flour -100 parts; Sugar -33 kg; Salt - 1.1kg; SMS - 4.2g; SMP - 1.5kg; Shortening - 24kg ; Invert syrup - 15 kg; Ammonium - 0.6; Water - 10 liter; Flavour – Vanilla

Preparation

1. Wheat flour is passed through a sifter removes all the dirt, stones etc.
2. Sugar is ground and fat is incorporated in molten form
3. Mixing: Ammonium bicarbonate, sugar syrup and water are mixed thoroughly in a high-speed mixer for a couple of minutes. Shortening and flavour are creamed for a few minutes. In dry mixing, maida, salt, sugar, SMS paste, SMP and vitamins premix are mixed. The mixing time is about 3–5 minutes.
4. Shaping and conveying to oven. The Rotary molder is used for shaping operation. This operation involved feed roll rubber roll and die roll and extraction belt and panning table belt.
5. Baking: The biscuits baked in an oven that has different temperature zone e.g. 120 °C, 350 °C and 150 °C.
6. Cooling: In cooling, two cooling conveyors are used. The cooling time is around 4 minutes.
7. Packing: The biscuits are packed in BOPP or any other moisture proof packaging materials.

(ii) Hard Sweet and Semi-sweet

All these biscuits are characterized by doughs that contain a well-developed gluten network but with increasing amounts of sugar and fat, the gluten becomes less elastic and more extensible. The prime requirement is that the biscuit should have a smooth surface, which has a slight shine or sheen and an even texture giving a bite that ranges from hard to delicate. These biscuits are commonly produced in many countries, particularly developing countries where the low cost of the formulation is attractive.

Early hard sweet types were with little or no sugar. The majority of popular types now available, such as Osborne, Marie, Rich Tea and Petit Beurre, all have very similar recipes and differ principally in their shape and thickness. It is difficult to add any flavorful ingredients successfully so most have a basically mild vanilla flavor or a caramel buttery flavor derived either from the use of real butter or synthetic buttery flavours. All have some syrup or malt extract. The fruit sandwich of currents (known as Garibaldi) or small sultanas is an interesting and significant variation.

Ingredients and recipes

Cabin biscuits with low sugar and fat levels are popular in African countries where the biscuit industry is developing and low cost and minimum effects of fat deterioration during storage are required.

Most semi-sweet types, despite varying names and shapes, have similar recipes at the upper limits of the fat and sugar level. The dough is mixed to about 40°C so the physical quality of the fat is less critical than for short doughs, there is enough dough water to dissolve the sugar completely so the crystal size is largely unimportant. Soft flours are used in the production of semi-sweet, hard dough biscuits. Flour from harder wheats with higher protein levels give less extensible gluten and higher dough water requirement. This type of flour results in harder biscuits. Some improvement in the suitability of these flours can achieved by dilution with up to 10% of starch such as arrowroot, com starch or potato flour. The quality of gluten can be tempered by the use of SMS (Sodium Meta bisulphate). This has a mellowing *effect* on the gluten, which is brought about by sulphur dioxide.

Typical general semi-sweet biscuit recipes are given in Table 3. Many semi-sweet recipes use baking powder, which is a mixture of sodium bicarbonate and an acidulant like acid calcium phosphate or sodium pyrophosphate. Normally the level of sodium bicarbonate is adjusted to give a baked biscuit with a pH of about 7, but some consumers like higher pH levels.

Table 3: Recipes for typical semi-sweet types on the percent flour basis

	Marie	Rich Tea	Cabin
Flour (9% protein)	100	100	100
Sugar	19	25	10
Fat	13	20	5.0
Syrup and or Malt extract	2.0	4.0	2.0
Skimmed milk powder	1.7	1.4	—
Salt	1.0	1.0	0.80
Sodium bicarbonate	0.40	0.60	0.80
Lecethin	0.26	0.40	0.10
SMS	0.030	0.035	0.030
Ammonium bicarbonate	1.5	0.40	0.80
Water (approx.)	2.4	19	20

Dough Mixing

There are four basic requirements for the mixing of these doughs. The ingredients must be blended, the flour has to hydrate, the sugar must dissolve and the hydrated protein must be kneaded to produce the three-dimensional structured material known as gluten. The hydration of the flour and dissolution of the sugar are time dependent, the others are related to the design and speed of mixer. Normally, all the ingredients are placed in the mixer together before mixing begins. In some cases the fat, water and sugar are mixed initially to allow dissolution of the sugar and plasticizing of the fat. It is usually necessary to mix to higher dough temperatures in large universal mixers than in small ones in order to ensure that sufficient kneading is given to develop the gluten. The mixing to constant final dough temperature is a prime control parameter. It is recommended to aim for a final dough temperature of 40–42 °C where SMS is used and 44–46 °C in unsulphited doughs.

If the mixer does not produce an extensible dough by the time these temperatures are reached (with not less than 4 min of vigorous mixing action), the starting temperature of the blended should be reduced (usually using cooler water) to allow an extension of the mixing time before the final temperature is used. The length of mixing time varies with the type and size of mixer and the level of SMS used. Slow-acting vertical spindle type mixers may require as much as 50 min, speed of around 60 rpm require about 20-25 min. and small mixers with dough capacity of around 160 kg and a beater speed of 90 rpm can mix satisfactory dough in 4.5min, use of SMS has a dramatic effect on the dough quality. The use of about 0.03 % of this salt per 100 parts of flour allows at least a 10% reduction of dough water and a significant reduction of mixing time compared with doughs using no SMS. The benefit of SMS is that is acts immediately, in fact it can be added towards the end of dough mixing with the satisfactory result that the reaction goes to completion very quickly.

The enzyme proteinase will accelerate gluten softening, resulting in reduced mixing times. The quantity to use will depend upon the type of flour available, on the dough temperature, fat and sugar concentrations and on the activity or strength of the enzyme preparation

Dough piece forming

Dough sheeting is another very critical part of the process. The dough must be supplied in a suitable quality for sheeter to produce a continuous homogeneous sheet witha smooth surface. Thus dough of acceptable consistency and quality as judged at the end of mixing may produce inferior biscuits if the handling of that dough before sheeting is variable or careless.

Dough at about 40 °C should be protected from cooling used without delay and scrap dough should be immediately mixed with it in the sheeter as metered in some other way. It may be necessary to heat the steelwork of the pre-sheeter, sheeter, guage roller etc., to avoid chilling of the dough and the baker air should not be allowed to chill or skin the dough surface at least prior to sheeting. On standing, a semi-sweet dough tends to lose its extensibility but a small amount of remixing usually revives the dough. The standing time before use of a dough should not be more than about 3- min.

A three-roll sheeter is used to form a continuous sheet of dough. Holes or imperfections in the sheet will be lost as the dough is laminated. Following the sheeter there should be two or three pairs of gauge rolls before a sheet at the correct thickness is formed. Too great a reduction may cause distortion and damage to the dough structure, which will affect the development during baking and also the biscuit shape.

Sometimes the dough will tend to adhere to one or the other of the gauge rolls such that release is difficult and the dough surface is spoilt. This problem can be overcome by some techniques. An air blower to skin the dough before the gauge rolls is frequently sufficient but failing this or light flour dusting may be necessary.

Semi-sweet doughs exhibit a certain amount of elasticity so the correct biscuit shape can be achieved by allowing a relaxation of the dough prior to cutting. This is usually achieved by rippling the dough onto an intermediate web or onto the beginning of a long web prior to cutting. The relaxation results in some thickening of the dough sheet so the setting of the final gauge roller is always less than the thickness of the dough sheet at the cutter.

Semi-sweet biscuits are always cut with a complete surround of 'scrap' dough. This is lifted clear and returned to the sheeter for reincorporation. This dough is denser and often cooler than the fresh or virgin dough. Semi-sweet biscuits are always dockered and usually bear a name and stamped into the surface. With a reciprocating cutter this dockering and printing is made at the same time as the cut of outline. As the cutter rises an ejector plate rests on the dough surface to effect a clean release. After cutting the scrap is lifted away. Following the cutting operation the dough pieces may be garnished with sugar or other granular material or washed with milk or an egg/milk mixture to enhance the gloss and appearance after baking. Most semi-sweet biscuits are no~ garnished at all but if any sort of surface dressing is applied, care must be taken not to spill the dressing onto the panning web, otherwise either the oven band or the subsequent performance of the panning web will be impaired.

Baking

Lift, or the development of biscuit structure is a result of gas released from the leavening chemicals and the expansion of water vapour as the temperature rises. The biscuit can be up to 4–5 times thicker than the dough piece entering the oven and the moisture content is reduced from about 21 % to less than 1.5%. As moisture removal, to a relatively low level, is necessary to avoid the condition known as checking when the biscuits cools, it is normal to bake semi-sweet biscuits on wire type oven bands. However, Marie biscuits and sometimes-other thin types are traditionally baked on steel bands. A pre-requisite for semi-sweet biscuits is a smooth surface of even, fairly pale colour with a sheen. The smooth surface and even lift is determined by the condition of the dough surface after sheeting and gauging. The sheen can be enhanced by passing steam into front of the oven to increase the humidity of that section so that it exceeds the dew point at the dough surface. This allows a film of moisture to be deposited which becomes sheen when it dries out later in the oven. At the oven exit checks must be made to determine whether the biscuit size, shape, colour and moisture are within the limits set for quality and packet specifications. Colour measurement (the reflectance value or darkness) of the top surface of biscuits can be made continuously, however colour on both top and bottom surfaces are important and as yet the underside seem to be neglected.

Checking (cracks in the biscuits after packing) is a potential problem for semi-sweets biscuits. In order to prevent this it is necessary to bake to low overall moisture content or to cool the biscuits carefully.

Flavoring of biscuits

The large amount of moisture that is lost during baking of semi-sweet biscuits and their low final moisture content make it very difficult to flavor these biscuits. Volatile flavoring materials added to mix are driven off in the oven. One technique that may be considered is to oil spray with flavored oil. The biscuits must be hot enough to allow the oil to soak in but not so hot that the flavours in the oil are volatilized.

Cooling and handling of biscuits

Semi-sweet biscuits usually strip easily from the oven band because they are rigid even when hot. It is usual to let the biscuits cool in air before packaging. Cooling conveyors are typically two or three times as long as the oven and normally of the same width as the oven band. For high capacity plants very close attention is needed to the biscuit handling up to the wrapping machines so that the transfer can be made with minimal human intervention and breakage. A number of plants have forced cooling of biscuits, which have been stacked immediately after stripping from the oven band. This arrangement saves much space. Soft flours are used in the production of semi-sweet hard dough biscuits and frequently the flour is weakened by the addition of corn flour, or potato flour.

(iii) Continental Semi-Sweet Biscuits

Biscuits of this type are commonly made in France, Germany and Switzerland. The recipes are slightly higher in fat level and are mixed by two-stage process similar to short doughs. All ingredients except the flour are firstly mixed up to a homogeneous 'cream'. The dough is then rested for between 30 minutes and 90 minutes to reduce the stickiness, before sheeting and gauging. The formulation sometimes includes proteinase and this requires at least 60 minutes standing time for enzyme to react with the gluten. The dough is not normally laminated. These doughs tend to be sticky which makes them more difficult to process through gauge rolls and a good smooth surface may be achieved prior to cutting by ensuring that at least the final gauge is very clean and the dough pieces are often brushed with a milk wash to enhance surface appearance after baking. The resultant biscuits are softer and shorter in texture than traditional British semi-sweet types and surface is not as smooth.

(iv) Soft Dough Biscuits

Short doughs, which are soft enough to be just pourable, are called as soft doughs. Pieces are formed by extrusion in a similar way and in the same machine as wire cut and rout biscuits but nozzles rather die holes are used to channel the dough. The dough is pressed out either continuously or intermittently on the oven band that may be raised up and then dropped if discrete deposits are requires. As the band drops, the dough pieces break away from the nozzle. The biscuits produced in this way are usually rich in fat or based on egg whites whipped to a stable form, the dough must be very short to allow it to break away easily as it is pulled away from the nozzle.

The nozzles through which the dough is extruded are usually indented to give a pattern and relief to the deposits. Also by rotating the nozzles, swirls, circles and other attractive shapes can be developed. In the case Spritz biscuits, the nozzles are oscillated from side to side during continuous extrusion. This forms a ribbon of dough. Depositing allows not very fancy shapes to be formed, but also by synchronizing two or more depositors, different colored and flavoured doughs can be combined. Jams and Jelly can be added on the top of dough deposit.

Typical recipes for soft dough biscuits

Deposited biscuits

Ingredients	**I**	**II**
	(g)	(g)
Flour	100	100
Butler (salted)	54	70
Fine sugar	35	40
Fresh eggs	1	1
Sodium bicarbonate	0.2	1.0

Biscuits dust	20	
Salt	0.7	0.5
Water	7.5	6.0
Flavors	Yes	no
Invert syrup	1	—
Sodium pyrophosphate	1	—

Ingredients

Nearly all these biscuits are luxury types. The production rates are usually low and the ingredients are expensive. Butler is widely used also eggs, grounds almonds, coconut flour and cocoa. But contrast with wire cut types, coarse particle size ingredients are avoided as these block or interfere with the smooth functioning of the depositing nozzles. The consistency of the dough is critical so the temperature of the ingredients is important. A temperature of about 170 c is recommended for butter. The sugar should be fine or very fine as there little water is available to effect solution and fine crystals give a better eating texture in the baked biscuits. Some manufacturers maintain that ground biscuits crumb (from the same type of biscuits) aid the texture and structure. Care should be taken not to include crumbs from over baked product as it will adversely affect the flavor and colour of the baked product.

Dough mixing

As the dough of pourable consistency is required, detachable bowl type mixer is used. The mixing times are quite short and relatively gentle. It is usually best to cream up the butter with sugar, eggs, milk and water and to add the flour later with a minimum mixing to achieve a homogenous mash. Dough temperature is important to maintain consistency and correct fat dispersion. It may be necessary to cool the flour and certainly any water and milk and should be very cold. Dough temperature between 10-150 °C should be aimed at.

Dough piece forming

As already stated the dough is pressed out through nozzles on to the oven band which may be raised up and then dropped if discrete deposits. As the band drops pieces break away from the nozzle. The nozzles through which dough is extruded are usually required to give a pattern and relief to the deposits. Thus it may involve more than one type of nozzle on a single depositing machine located one after another or more commonly a series of machines located one after another and with synchronized action.

Baking

It is essential to use a steel band to bake products in this group. All types shows some spread but those rich in sugar spread the most. Fat rich products do not stick to the oven band. Treatment of the band with oil or flour may be necessary. Baking is normally slow at low temperatures. There is little water to remove so the baking process is principally to develop the texture and colour the surface. The letter may be very irregular with fine peaks, which will colour very easily if the oven temperatures are too high. The biscuits have a soft and "melt in the mouth" type and are very delicate and easily broken.

Biscuit Handling and Packaging

Where the biscuits are thick and irregular shape they do not bend themselves to stacking and mechanical handling into wrapping machine like other biscuits. It is, therefore, usual to transfer the pieces individually in to trays, boxes or tins prior to final packing.

(v) Sponge Batter Drop Biscuits

Recipe

Ingredients	(g)
Flour	100
Fat	3.2
Fresh eggs	6.5
Fine sugar	80
Sodium bicarbonate	0.14
Sodium phyrophosphate	0.20
Salt	0.80
Glycerine	3.0
Glucose syrup	6.2
Water	10.0

There are variations on the sponge mix recipe but in all cases, the dough is an aerated batter that is pumped to a sponge pipe depositor or in baking trays according to the set routine and at end of each deposit the holes are shut off to prevent drips. It is important that the batter is not too stringy otherwise tails are formed at the end of each drop.

Sponge batter mixing and depositing

It is normal to make the mixture of batter in two distinct places. Firstly, a premix of all the ingredients (eggs. flour. sugar and water) is blended together to form a more or less homogenous slurry. This is then pumped for the aerator and metering pump that supplies the deposit manifold. Air is metered in and the batter is converted into fine foam. During aeration, it is necessary to provide cooling to prevent the overheating of the batter. The density should be about 0.88/C.C. and the temperature 190 °C. Depositing is with a depositor, that is a sponge pipe that follows the oven band during the depositing stage then moves back while holes in the pipe are closed.

Baking of sponge drops

Baking is usually in a moderately hot steel band oven for about 8 minutes. The fat less batters opt to stick badly to the oven band during baking so it is always necessary to grease the band in someway. Considerable difficult can be encountered in the search for the optimum means of preparing the oven band. There are various techniques but all involve the use of flour in addition to an oily lubricant, spreading the oil and flour evenly and at the desired low levels either together as slurry or separately, is a major engineering challenge. Without the flour the batter may spread to an unacceptable extent prior to setting in the oven, with insufficient oil baked pieces may adhere so firmly to the band that stripping is virtually impossible

(vi) Fermented soft dough biscuits

In India fermented dough biscuits are prepared by fermented slurry and then it is mixed with rest of the batch for dough mixing. Slurry is fermented for 2-3 hrs only. The composition of typical slurry is given below.

Recipe

Wheat flour	90 kg
Sugar	4kg
Yeast	21kg
Mixing time	2–3 minutes
Fermentation time	2 Hrs

Preparation of batch

Wheat flour	160 kg
Sugar	30 kg
Scrap	30 kg
Invert syrup	9 kg
Ammonium bicarbonate	8 kg
Salt	3.7 kg
Sodium bicarbonate	1 kg.
Sodium metabisulphite	100 ml
Lactic acid	450 ml
Mixing time	6–7 minutes

After preparation of the slurry, batch ingredients are added in the slurry and mixed for 6–7 minutes using vertical mixer. The dough is ready for further processing.

Shaping and cutting

A laminating machine helps to make 8–10 layers. This imports good puffing and crispness to the biscuits.

Sheeting

There is a cluttering roll and rubber roll on the lower side for putting up the pressure.

Baking

After cutting the biscuits are baked in an oven. The baking time is around 3–4 minutes.

Oil spray

To improve the shining of the biscuits coconut oil is sprayed on both sides of the biscuits which gives better appearance and eating quality to the biscuits.

Cooling

Cooling conveyors at room temperatures does the cooling of the biscuits. The cooling time is about 3 minutes. At the end of the cooling conveyer which detects metal pieces if contaminated from any part broken part.

Packing

The biscuits are packed in BOPP wrap and then sealed.

SPREAD OF THE BISCUITS

Spread of the finished biscuits is the most important character, which should be carefully controlled as its excessive variation may create serious problems in the product line.

Factors allowing greater spread	**Factors which reduce spread**
Factors related to flour in the formulation Coarse flour particles Minimum mixing after flour addition	Higher flour water absorption value, including heat treated and chlorinated flours Over mixing of dough
Factors related to sugar in the formulation Sugar with low means aperture Increased quantities of crystalline sugar	Sugar with high mean aperture size Lower level of sugar
Factors related to fat in the formulation Soft doughs due to higher temperature More fat	Cold doughs Less fat
Factor related to aeration in the formulation	High dough pH (more Ammonium or sodium bicarbonate)
Factors related to dough age and dough piece weight High dough piece weight	Old dough
Factors related to oven conditions Greasy oven band Low temperatures in front of oven	Flouring of oven band Higher bake temperatures, faster Baking

Wide variety of Biscuit textures

Wide variety of biscuit colour

Fig 4.2. Wide variety of biscuit textures and colours

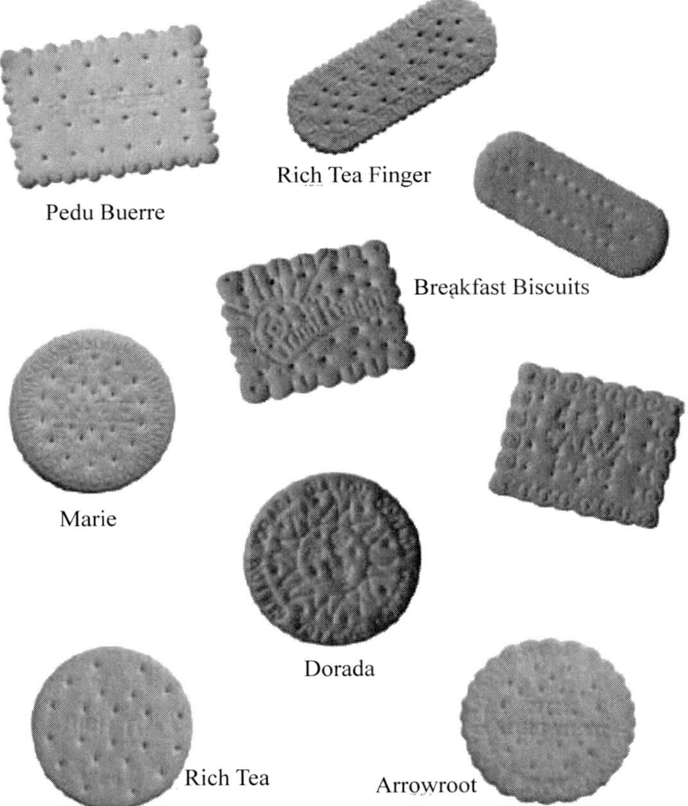

Fig 4.3. Different types of Biscuits

Processing Technology of Cookies

Wheat is considered unique as the basic raw material for the production of bakery products such as breads, biscuits and cookies because of its special gluten development properties as compared to those of other cereal grains. The wheat flour has added advantages of being relatively cheap, low moisture content, nutritious and its dough has very good machinability due to its elasticity as well as extensibility and hence suited to continuous commercial production.

After green revolution due to introduction of high yielding Mexican dwarf wheats in the late sixties, it has been observed that there is a galloping increase in the production of wheat in India. Because of the surplus and the availability of wheat ideally suited for the production of baked products, the bakery industry in India has made a significant progress during the last decade. Today, bakery industry has come to occupy an important place on the industrial map of the country. It has already recognized as the largest of the processed foods industries in the country; touching a level of annual consumer outlay of about ₹10,000 crores.

As compared to biscuits and crackers the cookies, in India, resemble home made baked products and commercial production of cookies in India is quite low. In recent years, however, an increased interest has been noted in the production of cookies, because of their attractive features like good catering quality and long shelf life. Good catering quality makes cookies a popular snack and long shelf life allows large scale production and wide distribution.

Classification of Cookies

Cookies are chemically leavened baked products with top surface broken by fairly wide cracks somewhat evenly spaced to give uniformly sized islands. They have richer crust colour and a moisture content ranging from 1–4 percent. Cookies differ from biscuits in respect to their crispness, bite, taste, texture and level of ingredients used. Cookies are generally crisper and contain larger amount of sugars and shortenings than biscuits.

Cookies can be classified into four major groups depending upon the kind of equipment used to form the individual places.

(a) Deposit Cookies
(b) Rotary-moulded cookies
(c) Wire-cut cookies and
(d) Cutting machine cookies

(A) Deposit Cookies

This category of cookies is made form very soft dough deposited directly onto the oven band by a forming machine. Deposit cookies contain about 35–45 percent sugar, 60–70 percent shortening and unbleached soft wheat flour with 8–8.5 percent proteins and 0.35–0.40 percent ash. The flour must be able to carry the sugar and shortening without too much spread, so that the top design is preserved through baking. At the same time, the flour and other ingredients must contribute enough adhesive properties to the dough so that it will adhere to the band and pull away form the main tube dough in the deposit stage.

(B) Rotary-Moulded Cookies

Rotary-moulded cookies are made from crumbly dough pressed into a form on a rotation cylinder, later removed and deposited onto the oven belt. Rotary moulded doughs are often high in sugar and fat but low in water content. Most manufacturers use flour of about 8.1–8.2 percent protein for rotary moulded cookies, although a range of 7.1–9.2 percent has been reported. Ash should be about 0.415 per cent, with a known range of 0.33–0.47 per cent being used satisfactorily. The dough is crumbly, lumpy and stiff with virtually no elasticity. Cohesiveness of dough is due to fat used. For rotary moulded cookies, the dough consistency must be such that it will feed uniformly and readily fill all of the crevices of the die cavity under the pressures existing in the feeding hopper. During baking, dough spread is minimum. Lecithin at about 0.4 per cent level is added to improve machinability.

(C) Wire-Cut Cookies

Wire-cut cookies are extruded products of slightly stiff dough extruded through a die and cut by an oscillating wire. It is necessary to have the wire cut dough sufficiently cohesive to hold together as it is extruded through an orifice, and yet it must be relatively non-sticky and short enough, so that it separates cleanly as it is cut by the wire. Formulae may contain up to 100% sugar and 100% shortening based on the flour weight. Advantages of the Wire-cut cookies over rotary moulded cookies are more open grain and softer texture, and, as compared to deposit cookies, a more uniformly shaped cookie. Disadvantages over the rotary moulded piece are the lack of potential for making a surface design and somewhat less uniformity of size and shape.

(D) Cutting Machine cookies

Cutting cookies are those cut into appropriate shapes from a sheeted cookie-dough. For cutting cookies, the dough must be properly developed to provide tensile strength and extensibility for sheeting. A dough with slightly less fat and sugar but more water is used than rotary moulded dough.

Table 4. Ingredients of cookie formulations

Type of cookies	Deposit	Rotary moulded	Wire-cut	Cutting
Ingredients				
Flour	100	100	100	100
Sugar	45	45	50	30
Shortening	45	63	50	15
Water	17	12	13	12
High-Fructose	—	—	3.5	8
Corn syrup				
Non-fat Dry Milk	2	2.8	1	1
Egg Albumin	—	1.12	-	3
Sodium Chloride	1.0	0.6	1.5	—
Sodium Bicarbonate	0.2	—	0.2	0.5
Calcium Phosphate	—	—	0.7	—
Ammonium Bicarbonate	—	—	0.7	—
Lecithin	—	—	0.4	—

Process for Cookie Making

Typical cookie making process can be described as follows.

A. Preparation of Ingredients
B. Mixing of Dough
C. Cutting and shaping the Dough
D. Baking
E. Cooling
F. Packaging

Preparation of Ingredients

Soft wheat flour with particles less than 38μ gives the most desirable cookies. Therefore, flour fraction with 38μ particle size should be used. All the ingredients should be weighed separately according to the recipe. Water requirement depends upon the recipe and it is never constant. It also depends on the flour quality as water retention capacity of flours varies due to varying degree of starch damage.

Mixing of Dough

The mixing requirements vary for different products and also depend upon the raw material. Cookie doughs are usually mixed in upright horizontal mixers, low speed and short time cycles are used for mixing cookie doughs, because gluten strength is neither necessary nor desirable for sweet doughs. Generally, two methods are followed for mixing of cookie dough.
1. Creaming method and 2. All-in-one method.

1. Creaming method

In this method first the sugar and shortenings are creamed and then syrups, eggs, mild and salt are added, water with leaving agents is added next, and, finally the flour is added and mixed to the proper stage. The mixing is carried out at slow or medium speed of about 25–35 rpm. The excessive speed of mixing raises the temperature of dough and the fat melts, which caused the stickiness of dough and cause problem in machinability. Therefore, mixing speed and time are important for good dough development. The beneficial effects of creaming, as opposed to other mixing method, lie in the fat-coating effect which delays the hydration of flour proteins and starch, and the incorporation of small air bubbles, which assists in leavening and establishing the structure of the finished product.

2. "All in one" method

It involves the mixing of all the ingredients in a single step. This method is simple and easy in operation. All in one mixing facilitates better dissolution of sugars in the dough.

Cutting and Shaping the Dough

There are three general methods of forming of shaping cookie dough.

1. Pressing the dough into a die cavity and extracting it onto a moving belt. The dough to be formed into die cavity should form a solid lump when pressed together, but should possess little or no elasticity and have the general appearance and texture of crumbly lumps of shortening and sugar. Furthermore, they must possess sufficient cohesiveness so that the individual cookies do not tear apart at transfer points.
2. Extruding the dough or batter, which may be formed in fancy shapes by moving the orifice, and which may either be cut off by an oscillating wire or deposited on the moving oven belt without cutting.
3. Cutting shapes from sheets of dough, either with or without docking or embossing a design on the surface.

If the dough is to be formed into a sheet to maintain its continuity and uniform thickness so that it does not tear. It must have a certain amount of elasticity and adequate tensile strength to bear its own weight. Such dough is, as a rule, not suitable for forming into die cavities since they do not usually fill the die completely, they may not cut off cleanly, and they tend to shrink, distorting the design. Conversely, the rather soft dough suitable for extrusion cannot be extracted from the dies properly.

Baking

Cookies are generally baked in traveling belt oven. The dough pieces are continuously fed on oven band, the speed of the bank can be adjusted to suit the baking time for different types of cookies. The ovens are generally, divided into three zones. The recommended temperature pattern for most cookies is a fairly low temperature (150–165 °C) in the first zone, where the fat melts and undissolved sugars and chemicals pass into solution, and the whole of the cookie piece becomes soft and spreading of cookie dough takes place.

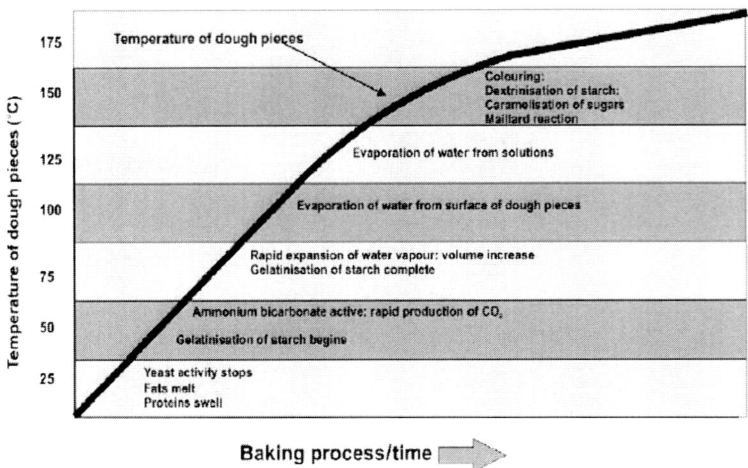

Fig 4.4 Graph showing relation between Baking time and Temperature

The chemicals used produce gas, and the heat causes it and the air already present, to expand, resulting in the cookie increasing in volume. A considerable high temperature (200–205 °C) in the intermediate zone is provided where setting and baking of cookie takes place due to coagulation of proteins and gelatinization of starch. In final zone slightly lower temperature is recommended to give desired colour and flavour to the cookies. The baking time of cookies is recommended between 10–15 minutes. Because after 15 minutes of baking the cookie diameter becomes constant and, the loss of volatile matters increases.

Table 5. Effect of baking time on cookie diameter and loss of volatile matter

Baking time (Min)	Cookie diameter (Cm)	Loss of Volatile matter (%)
0	7.0	0
4	8.5	6
8	9.5	12
12	11.5	26
15	12.5	40
18	12.5	70
20	12.5	85

Cooling

Hot cookies must be cooled uniformly before they are packed or sent for any secondary treatment. Non uniform cooling may lead to varying stresses at different points of the product, which will result in cracking and hence undesirable. The cooling is achieved by transferring the cookies in a single layer to a canvas conveyor and allowing them to travel on the belt for double the baking time.

Packaging

The cookies have the following quality characteristics those affect the packaging and shelf life of the product.

1. Low moisture content and hence the product has to be protected from moisture pickup during storage.
2. The product has crispness and therefore, it is brittle and hence should be protected from breakage during handling and transport.
3. The product is rich in fat and thus it should be protected from air to prevent development of fat rancidity and off flavour.

Keeping in view the above properties, packaging of cookies is done in unit packs consisting of paper, aluminum foil and polyethylene. Corrugated fiberboard boxes are used for bulk handling. These boxes are used to contain the small packs.

Role of Major Ingredients

Wheat Flour

Flour constitutes the primary raw material to which all cookies formulations are related. It provides a matrix around which other toughening or tenderizing ingredients in varying proportions are mixed to form dough or batter systems. A distinguishing characteristic of cookie flour is that, it is relatively coarse as compared to cake flour; however, it is finer in granulation than hard wheat fours, which are used for the preparation of bread.

The raw material of foremost importance in cookie making is the wheat flour. The flour obtained by milling in roller flour mill with 70–72 percent extraction is preferred.

(a) **Flour quality** is defined as "the ability of the flour to produce an attractive and product at competitive cost, under conditions imposed by the end product manufacturing unit". The tests most commonly used to characterize cookie flours and described as follows.

(b) **Protein content and flour strength**: the quantity and quality of protein present in flour is considered important in dough making. Cookies require a softer type of flour, which provides for structure building and leavening. The quantity and quality of gluten in flour influence the flour strength. It is observed that the stronger flour yields harder cookies with lesser spread and more puffing in the centre. Standard cookies flour is generally soft flour having protein content between 7.0–8.0 per cent.

(c) **Viscosity**: The protein strength is often measured by the viscosity method, in which the flour viscosity is noted in dilute acid solution. In a dilute lactic acid solution, flour gluten swells considerably and starch to a limited extent, this increases the viscosity of flour-water suspension. The increase has indirect relation to the swelling properties and quantity of gluten present in the total flour.

(d) **Ash content**: Ash colour is closely related to the colour-influencing components such as bran of flour. The bran, outer covering and aleurone layer have higher ash content than endosperm in wheat.

(e) **Particle size**: Flour granulation is of utmost importance in cookie baking. Different fractions of air-classified flour have different baking properties. Finer fractions of flour have low protein content and give greater cookie spread, than the coarse fractions. It has been observed that soft wheat flour fraction containing particle size smaller than 38 microns gives the most desirable coolies in relation to parent flour. Large particle size fractions give poorer cookies.

(f) **Colour of flour**: The flours best suited for cookie purpose will have a slightly yellowish tine due to the presence of natural pigments in the endosperm. The colour of the flour decreases with the storage due to oxidation reactions. Cookie flours are not improved by oxidizing treatment and this processing step is generally avoided in making the cookies. Increasing the intensity of chlorine treatment of soft wheat flour reduces the diameter and increases the thickness of test bake.

(g) **Alkaline water retention capacity (HWRC)**: The results of alkaline water retention capacity directly correlate with cookie diameter, since the test is done at pH 8.0–8.1, the conditions that actually exist in cookie doughs. It is much more informative, when combined with baking.

(h) **Damaged starch, Diastatic activity and Maltose value**: Damaged starch is one, which has been physically damaged during the milling process. Mechanical injury to the starch makes it more susceptible to enzyme action, which results of differences in intensity of grinding during production of flour. Granulation and its relationship to starch damage both must be considered to properly assess the effects of granulation in cookie spread. Both factors are dependent on grinding practices, severity of grinding and type of wheat relative hardness being milled. The diastatic activity is the capacity of the flour to produce sugars i.e., its amylolytic activity. For cookie making, high diastatic activity is not desirable and the flour unfit for bread-making purposes due to low diastatic activity can be easily used for cookie making. Maltose value of 2.0 (200 mg of maltose per 10 g flour) is considered as maximum for the production of cookies. The diastatic activity is the test, which reveals the extent to which the diastatic enzymes alpha-and beta-amylases produce sugars while acting on starch present in the flour. Normally, wheats have sufficient beta-amylase activity but lack in alpha-amylase activity. However, amylase activity increased thousand folds during wet harvest or germination. The diastatic activity is expressed as mg maltose produced/10 g of flour.

(i) **Banking test-spread ratio of cookies**: No single test can be of much use unless the actual product is made and hence the cookie-spread factor of flour is of great importance. The control of cookie spread is one of the most serious problems faced by production supervisors. Minor variations in appearance, flavour and texture are usually accepted with little complaint, but a cookie which spreads so much that it cannot be filled in the package, or one that spreads too little, causing slack fill or excess height for the package, can create have on the packaging line and generate large amounts of scrap. The width (W) and thickness (T) of the cookies are measured and spread ratio (W/T) is calculated after baking. High (> 9.0) W/T ratios are desirable characteristics of the flours for preparing cookies.

(j) **Specifications for cookie flour**: Soft wheat flour is particularly good for making cookies. Generally, soft wheat flour has relatively lower protein content, more mellowed gluten quality, lower absorption capacity, less starch damage, more spread and relatively short mixing tolerance properties as compared to hard wheat flour.

Soft wheat flour, as illustrated by the farinograms, develops and breakdown faster than hard wheat flours. Soft wheat flours absorption is lower than HRW flours (Hard red winter wheat flours). The weak dough and low absorption of soft red winter wheat flour (SWR) most likely reflects the differences in gluten quality and quantity between hard and soft wheat flours. Because of these differences, together with the others as mentioned before, soft wheat flour is superior to hard wheat for making sugar cookies by the better spread ratio and top grain with soft wheat than with hard wheat flours.

a. Cookie from soft wheat b. Cookie from hard wheat

Fig 4.5. Cookies from different types of Wheat

Water

Water is an essential ingredient in cookie dough which when added to wheat flour helps in the formation of gluten by hydrating the gluten proteins when dough is subjected to mechanical mixing. The gluten provides to the dough its characteristic rheological properties like desired strength, extensibility, gas retention power and elasticity.

The water has additional functions as follows:

1. It helps in maintenance of a particular temperature of dough during mixing.
2. Amount of water determines the consistency of dough.
3. It helps in the distribution of dissolved salt and sugar uniformly throughout the dough.
4. It facilitates the activities of enzymes in dough.
5. Water helps in the gelatinisation of starch during baking.

It is the quality of water which is important in cookie making. Microbiologically water used for doughs should be as free from microorganisms as is necessary for drinking water. Infected water may have a deleterious effect on human health even when the microorganisms are destroyed during baking. Hardness of water also affects the dough characteristics. Soft and hard water are the terms linked with the amount of soap needed to produce a lasting lather and this is particularly due to the levels of calcium and magnesium. Normally water with medium hardness (50-1000 ppm) with neutral pH is preferred for cookie making. Water that is too soft can result in sticky dough because of the absence of gluten tightening minerals. On the other hand water that is too hard results in very tough dough which prevents the spreading of cookies during baking.

Certain metals, particularly copper and iron, have marked catalytic effects on the development of rancidity in facts and oils. Drinking water standards demand low concentrations of copper and other metals associated with the development of fat rancidity, so this problem is likely to be under control if only drinking water is used in dough. If, however, it is suspected that metal ions are the cause of difficulties in dough quality, it is possible to reduce their effects with chelating agents like EDTA. In conclusion it is felt to be wise and good practice to select water for cookie making that is of constant quality and which confirms to the international standard for drinking water.

Sweetening Agents

Sweeteners are regarded as the most important class of ingredients in cookies. Some unusual varieties can be prepared without flour, a few without water (as such), and number without added leaveners, but no cookie formula is possible without some form of sweeteners.

The primary purpose of sweetening agent is to make product sweet. The quantity of sweetening agent added is usually such that it has significant effect on the texture and appearance of the product as well as on favor. Machining properties closely related to the dough piece to oven conditions are also closely related to the type and quantity of sweetening agent employed. The sweetening agents used in cookie making have varied functions and may be divided into three categories:

1. Sucrose and invert sugar
2. Derivatives of cornstarch
3. Other sweeteners.

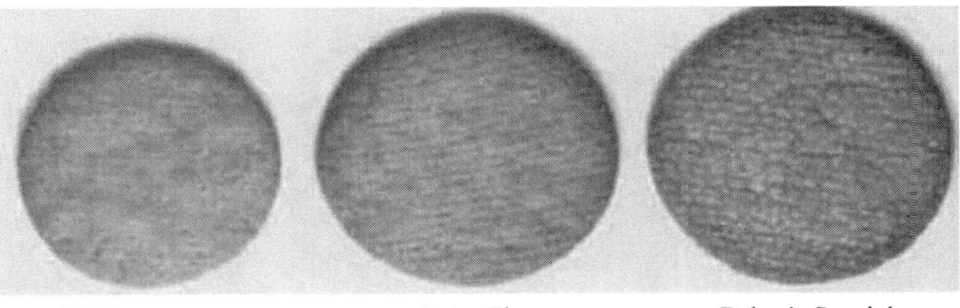

Medium Extra Fine Baker's Special

Fig 4.6 Cookies made with different granular sizes

Shortening Agents

Any edible fat used in bakery products is known as shortening. Shortening is essential components of most cookies. The amount of shortening in the formula influences both the machining response of the dough and the quality of the finished products. The saturated fatty acids are more important than unsaturated fatty acids as shortening in cookie making because the saturated fatty acids are chemically complete and stable and therefore, do not undergo much bio-chemical when stored.

1. Properties of shortenings

Shortenings should have a plastic nature over a wide range of temperature. Temperature plays an important role in the distribution of fat. If the temperature rise is higher than the maximum of the shortenings plastic range, then liquid oil will result, causing an oily dough while low temperatures tend to cause hardening of the shortening, causing uneven distribution in dough. The plasticity of shortening while mixing dough encourages the entrapment and retention of considerable quantities of air and thus contributes to the texture of the baked products. Further, the smaller the crystal sizes of the glycerides in a well plasticized fat the required plastic range, the greater the value in cookie making. Shortening, super cooled badly, having a large crystal structure gives dough with poor moulding potential and variable cookie weights and dimensions. Hydrogenated oils with their mono-diglyceride fractions encourage not only emulsification but also the homogeneous distribution of the fat soluble and emulsified ingredients throughout the dough and hence contribute to tenderness in cookies.

During hydrogenation process, hydrogen is added directly to the points of unsaturation in the fatty acids to convert the oils into solid fats, by which the stability of fat is increased to oxidative rancidity.

The "off flavour" developed in cookie during an extended shelf life is due to rancidity developed in the shortenings. These are mainly due to:

(a) Breakdown of fatty acid chains by oxidation,

(b) Spoilage by micro-organisms,

(c) Fat splitting by enzymes, particularly lipase and

(d) Absorption of foreign odours.

2. Functions of shortenings in cookie Dough

(a) Shortening reduces the toughness of dough

As gluten does not form until the flour is in contact with water and mixing action, the inclusion of fat tends to insulate and the gluten forming proteins from the water and consequently, a less tough dough results which is rather more extensible and ideally suited for cookie making. The greater the amount of fat, the greater will be the insulating effect. Excessive mixing will breakdown the insulation and tough dough will result again.

(b) It improves dough for machining and sheeting by lubricating the gluten.

(c) Controls the flow of cookies.

(d) Gives shorter bite to the goods

(e) Enhances the cookie flavours

Antioxidants

High level of fats may cause rancidity problem in cookies because of their long storage period. Therefore, use of antioxidants is very important in cookies to prolong their shelf life. Antioxidants are those compounds, which function by inhibiting the free redical mechanism of glyceride auto-oxidation and thereby retard the development of off flavour in products. At present time, only four chemical compounds are commercially important as antioxidants for foods. They are butylated hydroxy anisole (BHA), butylated hydroxytoluene (BHT), tertiary butyl hydroquinone (TBHQ) and propyl qallate. Synergists like citric acid or phosphoric acid may be added to improve the effectiveness of the antioxidants but they do not themselves function directly to prevent fat oxidation.

BHA is considered as the best synthetic antioxidant, because it has got very good carry-through properties and therefore, not destroyed during baking process, and hence extends the storage life of cookies and other baked products up to 3–4 times.

Leavening Agents

Leavening agents aerates a mixture and thereby lightens it. They also improve the texture and appearance of baked products. Leavening action may be produced by mechanical, chemical and biological means. In cookies leavening action is generally achieved by chemical and mechanical means. In cookies biological method of leavening is not practiced because the higher amount of sugars and shortening do not permit the efficient growth of yeast.

1. **Ammonium Bicarbonate:** Ammonium bicarbonate, often use in cookies, decomposes at high temperature into ammonia, carbon dioxide, and stem. Its usage increases spread and gives a large, more desirable "crack" in sugar cookies. But, it cannot be employed in moist, large volume bakery products as ammonia retention producers an objectionable strong pungent flavour and odour.
2. **Sodium Bicarbonate:** Sodium bicarbonates generate carbon dioxide and water in the oven by reaction with acids in the flour, leaving the sodium carbonate as the residual salt. Sodium carbonate has an unpleasant flavour and can react with fats to cause soapy tastes. Sodium carbonate has marked softening action on gluten causing spread and also darkening the product.
3. **Baking powder:** This leavening agent is produced by mixing an edible grade acid and sodium bicarbonates with or without starch or flour as

filler. Banking powder are classified as slow acting and fast acting. The fast-acting powders give off most of their gas volume during the first few minutes of contact with product. On the other hand, the slow-acting powders give up very little of their gas volume at low temperatures-they require the heat of the oven to react completely. Since banking time of cookie is short, therefore, it requires fast acting powder for better results. On the other hand cake where baking time is more requires slow-acting banking powder. Level of baking soda recommended for cookies is 0.4% on flour weight basis.

Fig 4.7 Different types of cookies

Soda Crackers

The term 'soda' or 'saltine' describes a very particular type of cracker. The soda cracker is an unsweetened, long fermented and laminated dough product. A typical soda cracker is a square biscuit approximately 50 × 50 mm with a thickness of 4 mm. Individual saltines are not formed and baked. Instead wide dough sheets are perforated by scrap less cutters prior to baking. After baking, these perforations form lines of weakness which enable the sheet to be broken into individual units. An additional feature of the product is the nine docker holes arranged in three rows of three which serve to tie the layers of dough together at those points. During baking, a good quality soda cracker puffs (springs) uniformly at every space between the docker holes as well as at the edge of the cracker. The top's blisters are uniformly brown while the bottom surface is nearly flat with many small blisters. The internal structure of the product consists of a series of layers between each docking hole generated by lamination during the manufacturing process. The cracker usually weighs 3–3.5 g and has a moisture content of 2.5%. The cracker is usually bland in flavour but with a unique crisp texture. The texture is the result of the laminar structure and low moisture content.

Soda crackers are made from dough that is lean relative to the other products of this category. A typical formulation has 8–10% shortening in the dough, up to 0–5% yeast, plus salt, and optionally, malt or malt syrup. The crackers are produced in a sponge and dough process with a lengthy sponge fermentation followed by neutralization with soda before sponge mixing and fermentation. The pH of the product does not drop appreciably during the dough fermentation, resulting in a slightly alkaline product; hence, the name 'soda' cracker.

Fig 4.8 Soda crackers

Cream Crackers

Cream crackers originated in the 1880s from an Irish firm named Jacobs. Although the product name implies that there is cream in the product, there is none. It seems that the name is traditional with no reference to the ingredients utilized to make the product. The cracker is similar to a soda cracker in that it is created from an unsweetened but long fermented, laminated dough. However, there are a significant number of differences between the two products. The cream cracker is usually relatively large (65X75 mm) and rectangular in shape. Its surface is pale with lightly browned blisters on both the top and bottom surfaces. The puffing and blistering give the product its uneven surfaces and a flaky layered structure that should be even throughout the interior. The finished moisture content is approximately 3–4%, slightly higher than for saltines.

All cream crackers have a simple formula containing flour, shortening (12–18%), salt (0.9–1.5%), water and yeast (1.0–2.4%). The dough is mixed in a single stage and fermented for a length of time defined by the manufacturer, which ranges from 4–16 h.

As with saltine the product is laminated, but in this case a fine cracker dust is added between layers prior to cutting and baking. The cracker dust filling, which consists only of flour, shortening and salt, is thought to facilitate separation between the layers of rather wet dough (approximately 26% moisture) during the processing. During baking, the laminations lift apart, form the irregular layers and give rise to the characteristic blisters and flaky structure. As is true for the production of saltines, a very hot oven is preferred to provide rapid expansion from steam and to dry the product.

Fig 4.9. Crackers being cut into strips

The texture of a cream cracker should be soft so that it melts in the mouth and does not shatter. The texture is a result of both the fat content and the degree of separation of the layers. Because there is no chemical leavening, the product's flavour is bland and slightly nutty.

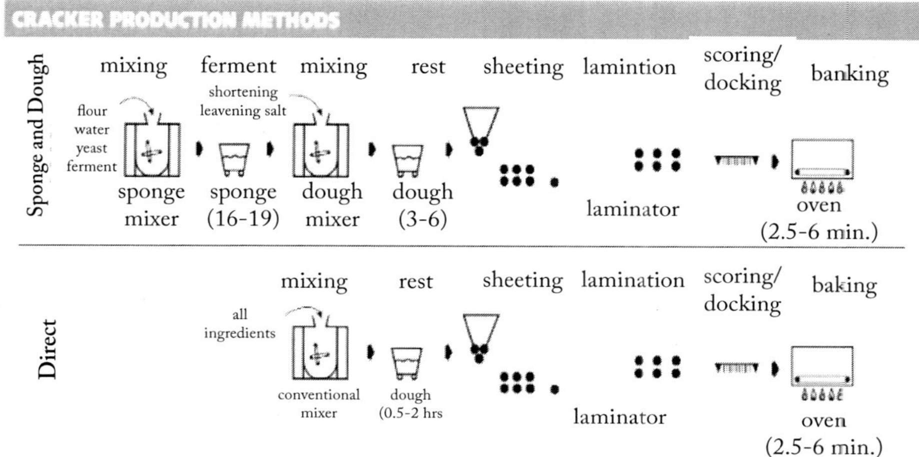

Fig 4.10 Cracker production methods

Snack Crackers

This group of biscuits may also be termed savoury or fat sprayed crackers. They are made in a wide variety of sizes and shapes, but the essential feature that defines the group is that they are oil sprayed while still hot from the baking process. The products may also be salted or dusted with a flavoured powder after the oil spraying. The flavourings may range from herb or savoury to cheese powders.

The products in this group may be generated by a range of manufacturing methods. As a rule, the dough is usually not fermented although exceptions exist which employ a one or even two stage fermentation. Generally, those products that have been fermented are also laminated; products generated without fermentation may be laminated or simply sheeted and cut.

Depending on the process utilized to create the dough, the products are either yeast or chemically leavened with most being chemically leavened. The texture of the products in this group depends on the manufacturing process utilized and differs from that of either saltines or cream crackers. In general, they have a more dense structure than that of either saltines or cream crackers and a relatively soft bite. Snack crackers have a finished moisture content that should not exceed 2%. The flavor of the product comes primarily from the fat spray and the topping applied. Surface oil sprays improve the mouth feel and enhance the appearance. It is common for a small amount of sugar or syrup to be included in the formulation. The sweetener acts to reduce the dry mouth feel and also as a flavor enhancer.

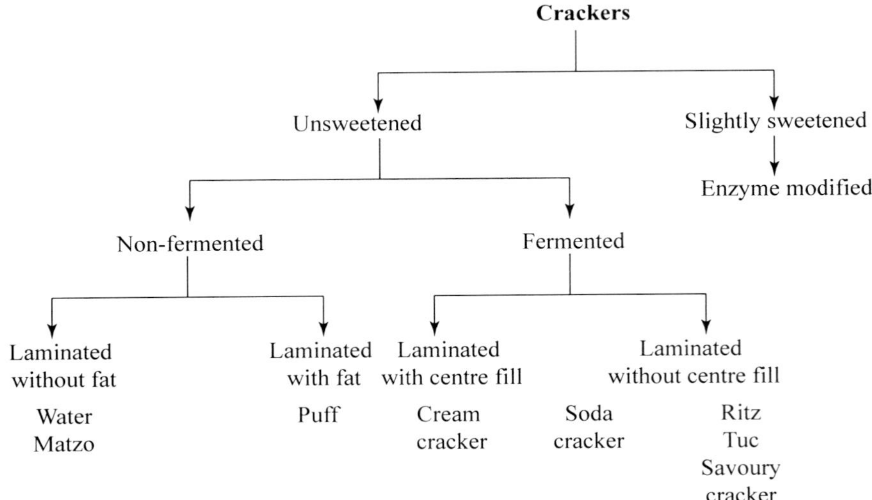

Fig 4.11 Classification of Crackers

In general cracker doughs are leavened and fermented with ingredients such as yeast, ammonia and sodium bicarbonate. Doughs generally have high water content (15–25%). Cracker doughs are usually laminated

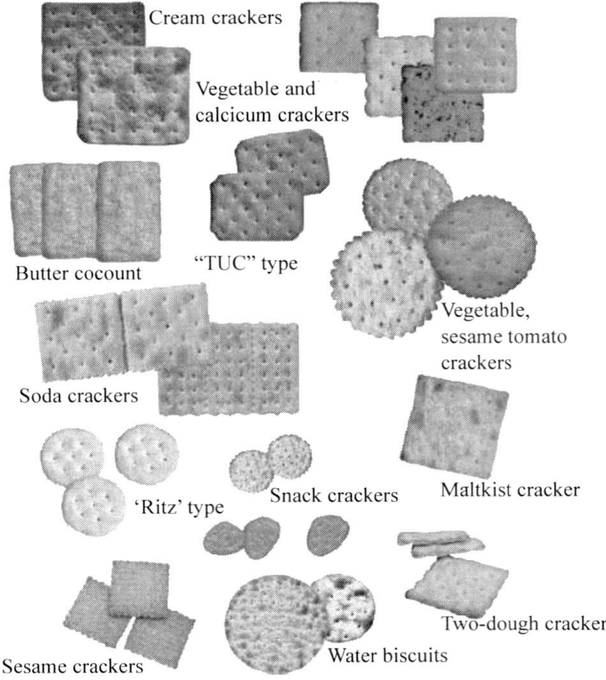

Fig 4.12 Different types of crackers

Manufacturing Technology

The manufacturing process used to produce all biscuits, coolies, and crackers consists of a mixing step, a shaping or forming step and a baking step. The mixing and baking steps are common to the manufacture of all types of these products. What is distinct for the products are the shaping or forming steps. The processing steps used to produce these products are as follows:

Mixing

Mixing is commonly defined as a process designed to blend separate materials into a uniform, homogeneous mixture. In the context of cookie and cracker dough the term takes on a broader meaning in that it also applies to the development of gluten from hydrated flour proteins, the aeration of a mass to give a lower density, and the dispersion of solids in liquids. One or more of the functions is required for the formation of cookie and cracker dough. These processes are accomplished with three principal types of mixers: vertical spindle mixers, horizontal drum mixers, and continuous mixers.

The Forming Process

While the same mixing and baking process may be used for many types of cookies and crackers, the forming step is specific to each product type. There are following three steps processes used to form cookie and cracker dough:

(1) Sheeting and Cutting

(2) Rotary moulding

(3) Extrusion

For each of these methods the rheology of the dough is different, and designed to be compatible with the process. In general, dough that are to be sheeted possess a significant gluten network as a result of mixing, and are both elastic and extensible. Those destined for rotary moulding lack gluten development and are best described as cohesive. Dough intended for extrusion are soft, frequently high in shortening, and spread while baking.

Sheeting and Cutting

The most common and versatile method to form cookie and cracker dough is by sheeting and cutting. This method consists of the production of a thick sheet of dough, evenly reducing the thickness of the sheet, cutting out the desired shapes, and returning the scrap dough to be reincorporated either in the mixer or early in the sheeting process. This method is used for the production of cracker, semisweet biscuit, and selected soft dough. After mixing, the dough is fed into a hopper, below which lie the sheeting rollers. There typically are three rollers below the hopper arranged in a triangular fashion. At least one of the top two rollers known as forcing rollers, is grooved so that a positive feed is provided to the gauge or gauging roller. The gauging roller, which is always smooth, serves to deliver the dough to the conveyor belt. The purpose of the sheeting unit is to compact the mass from dough hopper uniformly and provide a sheet of even thickness having the width of the processing line.

The relatively thick dough slab form the sheeter then passes through a series of reduction or gauge rollers. These are smooth steel rollers used to reduce the dough sheet to the thickness which is desired before cutting of the finished dough piece. The gauge rollers occur in pairs mounted vertically. For products having sticky or adherent dough it may be necessary to mount a scraper blade against one or both of the rollers to release the sheet of dough. On most process lines there are two to three pairs of rollers. This ensures that the thickness is reduced no more than 50% at any one rolling operation. Some dough, such as those of saltines and cream crackers, are laminated before cutting.

The lamination occurs by lapping the dough back upon itself in the process direction. At the lapper, the take away conveyor lies at a 900 angle relative to the line delivering the dough. The number of layers is controlled by the relative rate of the lapper and takes away conveyor. The lapped dough then passes through several more sets of gauging rollers to bring the dough sheet to the desired thickness prior to cutting. The repeated working of the dough in one direction results in an accumulation of stress. If the dough was cut at this point the resulting pieces would shrink to relieve the stress and misshapen or distorted products would result. Therefore, it is normal to relax the dough after reduction and before cutting. The relaxation is accomplished by transferring the dough to a conveyor, still moving in the same direction, but at a slower speed.

Once the dough has been relaxed it passes onto the cutting operation. Two different types of cutting methods exist: reciprocating cutters and rotary cutters. The reciprocating cutters are heavy block cutters that stamp out one or more pieces at a time. The cutter head may have a dual action whereby the cutter drops first, followed by a docking head or an embossing plate. The equipment operates via a swinging mechanism so that the dough sheet moves at a constant speed, the cutter drops and moves with the dough, then it rises and swings back to the original position. The second type of cutter, the rotary cutter, consists of a rotating metal cylinder. On the face of the roll are formed the desired shapes with a sharp metal edge. As the cutter rotates with the dough conveyor, the metal edges cut into the dough sheet to form the product. The product pieces are then conveyed into the oven.

As a result of either cutting process, from 20 to 60% of the dough sheet remains as scrap. The scrap dough is lifted way from the cut dough pieces and returned either to the mixer or to the sheeter. Return to the mixer permits uniform incorporation of the scrap into the dough mass. However, most systems route the scrap back into the sheeter either along the full length of the hopper or at the backside of the hopper. If dough is incorporated behind the new dough, imperfections will be on the bottom side of the dough sheet and will not be visible on the finished product.

Rotary Moulding

Three rollers are placed in a triangular arrangement below a dough hopper. A roller called the forcing or feed roller has deep grooves designed to pull dough down from the hopper. The dough is forced into the cavities of the engraved roller by the forcing roller. A scraper blade is mounted against the engraved roller to remove any excess dough and return it to the hopper via the forcing roller. Beneath the engraved roller is a rubber covered extraction roller that serves to drive the take away belt. The extraction roller applies pressure to the engraved roller via the belt, causing the dough to adhere preferentially to the conveyor belt. Dough pieces are dropped from the take away belt into pans or directly onto the baking belt. The rotary moulding process is suitable only for dry, crumbly dough. This process offers advantages over sheeting and cutting in that there is no scrap to recycle, and there are very low labour requirements to run the process.

Fig 4.13 Rotary Moulding Roll

Extrusion

There are two types of devices used in the production of extruded cookies: wire cut machines and bar/rout press. Both systems are very similar in design. A hopper is placed over a system of two or three rollers that force dough into a pressure chamber. The rollers may run continuously or intermittently to force dough out of the pressure chamber at the die. For wire cut cookies, the dough is extruded through a row of dies and a wire or blade mounted on a frame moves through the dough just below the die nozzle outlet. The cut dough pieces then drop into a conveyor band for transport to the oven. The wire usually moves only in one direction through the dough, opposite that of the conveyor. The wire cut machines operate at rates of up to 100 strokes per minute. Difficulties encountered with this type of production are distortion of the extruded dough piece during cutting, and inconsistent placement or drop of the cut piece onto the conveyor. The design of the bar or rout press is very similar to the wire cut machine. The hopper rollers and pressure chamber are essentially identical to their wire cut counter parts.

Unlike the wire cut machine, the base of the pressure chamber has a die plate that is inclined in the direction of the extrusion. A continuous ribbon of dough is extruded from a nozzle which is shaped to impart the desired finished product design. The dough ribbon can be cut into individual pieces by a vertically operating guillotine before the oven or after baking. If the product can be baked as a continuous ribbon the dough is extruded directly onto the oven band, otherwise it is extruded and cut onto a conveyor belt.

It is possible also to co-extrude two (or three) doughs one within the other. The central material may be a fruit paste or jam. Usually the co-extrusions pass as ropes and are cut by guillotine before or after baking. In order to make dough pieces where the inner filling is completely contained within the outer dough, there are special co-extruders that cut the extruded dough by a process called "encrusting"

Depositing

A very soft or pourable dough (or batter) is extruded through dies and deposited directly onto the baking band. Typically the depositing machine moves in synchronisation with the baking band while the deposit is being extruded, then rises and moves back as the extrusion is stopped. In addition, the dies may rotate to make the deposits swirls or twists.

Baking

The cookie and cracker industry relies almost exclusively on band or traveling ovens to bake its products. The band oven is essentially an insulated, heated tunnel equipped with a continuous conveyor. The ovens vary both in length (from 30–150 m) and in band width (from 1.0–1.5 m). More modern ovens frequently consist of a series of modular units or zones. Each of the zones is equipped with its own set of controls so that the temperature and air flow may be controlled within that zone. The oven band is typically continuous, passing onto a drive drum at the end of the oven and returning underneath the baking chamber to a tension drum at the feed or input end of the oven. The chamber through which the oven belt returns may or may not be enclosed. Frequently, the oven band serves as the baking surface for the product. Depending on the product type the oven band may be solid or any of a variety of open wire mesh types. Choice of mesh is a critical factor in the process as it affects the heat transfer at the bottom of the product. This, in turn, can have a marked effect on quality of the finished product. There are three basic typed of ovens: direct fired, indirect fired and fully indirect fired. Ovens are usually heated by the combustion of gas, although there are a few manufacturers who use oil or electricity for economic reasons.

The most common type is the direct-fired oven in which gas is burned inside the baking chamber itself. In these ovens, the burners are placed across the width of the oven at regular intervals, both above and below the oven band. In other oven types, termed 'indirect ovens', the gas or oil is burned outside the baking chamber and the heated combustion gases are circulated into and throughout the baking chamber. Indirect fired ovens typically have al single burner for each section. The hot gases from the burner pass along pipes parallel to the length of the oven, both above and below the oven band. The products of combustion are circulated throughout the baking chamber by large fans. Fully indirect ovens are those in which the heat source is independent from the baking chamber and heat transfer occurs via a heat exchanger. None of the products of combustion circulate inside the baking chamber. This type of oven is not common except when oil is used as a combustible material. If circulated, the products of this type combustion would impart an unacceptable flavour to the products.

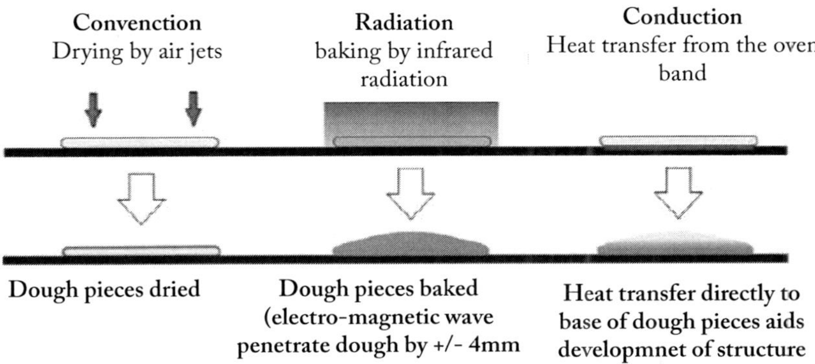

Fig 4.14 Baking stages

The structure of biscuits, cookies and crackers is developed in three stages. In the first stage ammonia, carbon dioxide gases and water vapour are formed and released. These changes cause dimension development and more lift in cookies and crackers. In the second stage - moisture is removed from dough piece, maximum gas expansion is achieved, starch gelatinization takes place, gluten proteins denature and crust surface begins to form. In stage three- structure is fully set, flavour development, colour starts forming sugar caramelization and maillard browning

Cooling

Products hot from the oven must be cooled prior to packaging for several reasons: the products may not be firm enough to withstand the packaging process while warm, the packaging material may shrink around a warm product, or the quality of the products would deteriorate if palletized while warm because the cooling rate across the pallet would be quite slow. The normal method of cooling products is to place them on an open conveyor and transfer them a distance 1.5-2 times the length of the oven. The products cool naturally in the ambient factory atmosphere. In a few cases, it is necessary to provide forced air to aid the cooling process.

Enrichment of the formula

All biscuits, cookies and crackers have flour as a major ingredient. There is usually some fat and sugar in their formulation and as the ratios of fat and sugar to flour are increased, both the method of manufacture and the eating quality of the baked product changes.

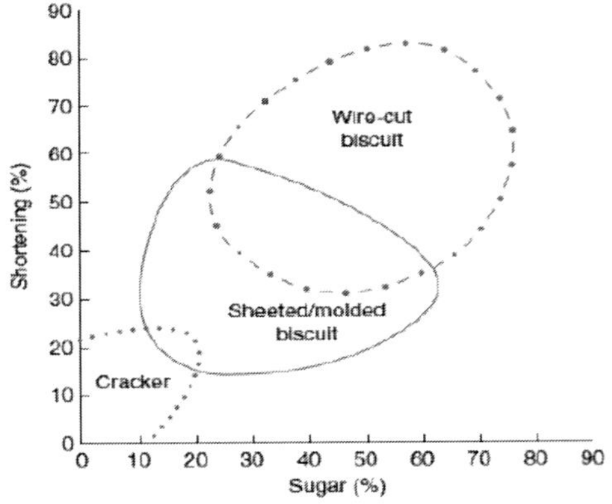

Fig 4.15 Illustration showing difference between cracker, moulded biscuit and wire cut biscuit

Secondary Processing

Sandwiching with cream

A deposit of fat/sugar cream is placed on a thin biscuit and then a second biscuit is placed on top and pressed down. The cream is then cooled and it sets firm and adheres the two shells together.

Cream sandwiched biscuits are very popular as the cream affords an ideal means of introducing flavours that would not be possible from baked products. The cream may be either sweet or savoury. In the case of savoury (salty flavours) the cream is a mixture of fat and a non sweet material like milk powder or milled biscuit crumb.

Chocolate coating (enrobing)

Chocolate and biscuit (or creamed wafer pieces) are an excellent combination and are very popular. Liquid and tempered chocolate (prepared with the ideal content of crystals to ensure a hard glossy chocolate when it is cooled) is applied either on one side or all over the biscuit. If only one side is chocolate coated the biscuit is said to be half coated, if all over it is fully coated. The coating may be either milk chocolate, plain chocolate, white chocolate or a chocolate flavoured coating.

Chocolate moulding

Biscuit or creamed wafer pieces are placed in moulds and the moulds are filled with chocolate. The chocolate is then cooled and knocked out of the mould. The moulded products look like chocolate bars.

Icing

The icing mixture is a suspension of very fine sugar crystals in water with a setting agent such as pectin, albumin or gelatine. The mixture is usually coloured and flavoured. Biscuits are coated (on only one side) with this mixture in the same way as chocolate is enrobed. The icing mixture is then dried in a warm tunnel and sets very hard. In another form of icing the icing mixture is foamed into a formable state and this is deposited in peaks onto the surface of small biscuits which are then dried.

References

Biscuit Cookie and Cracker Manufacturing Manual 1 Ingredients, Woodhead Publishing, Cambridge, 1998.

Biscuit Cookie and Cracker Manufacturing Manual 2 Biscuit Doughs, Woodhead Publishing, Cambridge, 1998.

Biscuit Cookie and Cracker Manufacturing Manual 3 Biscuit Dough Piece Forming, Woodhead Publishing, Cambridge, 1998.

Biscuit Cookie and Cracker Manufacturing Manual 4 Baking and Cooling of Biscuits Woodhead Publishing, Cambridge, 1998.

Davidson, I. Biscuit Baking Technology.2nd Edition, 2016

http://www.thebiscuitdoctor.com/about

5
Formulations and Processing of Cakes and Types of Cakes

Cakes are sweet baked products characterized by high level of sugar in the formulation and hence are foods of high calorific value. Among the various ingredient used flour, sugar, shortening and egg are the essential ingredients for cake manufacture. The optional ingredients are baking powder, milk, fruit etc. For cake manufacture the selection of ingredients is of paramount importance.

Ingredients Used in Cake Making

Ingredients are classified as follows:

The above ingredients are also classified as follows based on their function:

- Structure builders : Egg, Flour and milk.
- Tenderizer : Fat, Sugar and baking powder.
- Moisturizer : Milk, eggs, Honey, syrup

Flour

It builds the structure of the cake and holds other ingredients together in an evenly distributed condition.

- Flour for cake making should have a protein content of 7–9%.
- Flour from soft wheat is ideal for cake making.
- For cake making, flour should have fine granulation which affects the finesse of grain of cake.
- Cake flours are bleached. Bleached flour has low pH. Starch gelatinizes at a low pH and thus effects the faster setting of cake.
- Cakes made from strong flour will peak at the centre and will be tough and dry.
- Too weak flour will sink at the centre

Sugar

- Sucrose is the most commonly used sweetening agent in cake making.
- Due to tenderizing action low flour proteins, it makes cake tender.
- Being hydroscopic in nature, sugar helps to retain moisture in cakes and improves shelf life.
- Sugar has a lubricating action on gluten and thus helps in the process of acquiring volume in breads.
- The golden brown crust colour is due to the caramelisation of sugar.
- Dextrose- mono- hydrate is used to cut down the sweetness of cake. It contains 8% of moisture which should be adjusted in cake formula.
- Granulation plays an important role in any type of the sugar used. Large granular sugar will escape the aeration; even the small granular sugar will not be so desirable. Thus coarsely powdered sugar is best suitable.
- Liquid sugars (Molasses, Honey, Invert sugar, and malt syrup) can be used apart from powdered sugar. They have better water retention capacity and improve the shelf life of the product. They also impart good flavour and improve crust colour of the cake.

Shortening

- Fats have a tenderizing action on protein and make the cake tender.
- Fat plays an important role in holding air cells incorporated during creaming operation. These air cells have a tenderizing action on cakes.
- As a moisture retainer, fat helps to keep the cakes moist thus improving the shelf life.
- Fats used in cake making should be plastic in nature, which could hold minute air cells during creaming action.
- Granular fats should be avoided because they have poor capacity for holding aeration.
- Fats like hydrogenated shortening, butter, margarine are used in order to acquire specific characteristics in cake.
- Shortening should be able to maintain plasticity at room temperature.
- They should not melt by the heat produced due to friction during the creaming.
- Very hard shortening will not cream well and too soft or liquid shortening will not retain aeration

Egg

- Egg provides structure to the cake.
- Eggs are not aerating agents; but act as aerating agents because of air incorporated during whipping.
- Eggs provide moisture to the cake.
- Lecithin present in the egg yolk acts as an emulsifier and lutein, also found in yolk, imparts colour to cake.
- Eggs improve the taste, flavour as well as nutritional value.

Milk

- Milk solids perform the function of structure formation in cakes.
- Milk proteins have a binding action on flour proteins which may create toughness and dryness in cakes but this is prevented by egg, fat, sugar etc.
- Milk enriches the cake nutritionally.
- Lactose sugar present in milk improves the crust colour and moisture retention capacity of cakes.
- Milk also improves flavour and taste of the cakes.
- Apart from eggs, milk is the only other ingredient that provides moisture.

Water

- Water, whether added as such or in the form of milk, hydrates the flour protein and thus helps in the structure formation.
- Formation of gluten, release of CO_2 from baking powder and formation of vapour pressure are made possible by the pressure of water.
- Water helps to regulate the consistency of the batter which affects the volume and texture of cake.
- Shelf life of cakes is determined by the amount of moisture retained in the cake.

Salt

- Salt enhances the natural flavour of other ingredients used in cake making and thus improves the flavour of cakes.
- It improves the crust colour of the cake by lowering caramelisation of sugar.
- Should be used by dissolving in water so that impurities will not come into cake mixture.
- Salt helps in cutting down excessive sweetness in cake.
- A pinch of salt always improves the taste and flavour.

Flavouring agents

- Flavour is a very important aspect of quality product.
- A flavouring agents action should not be impaired due to heat or storage.
- Cheap flavours often break down under the influence of heat giving off flavour to the product.
- A flavouring material should be added after measuring, excess will spoil the gastronomic appeal of the product.

Baking of Cakes

- Different kinds of cakes are baked at different temperature; temperature is adjusted according to the richness of the formula.
- Richer the cake, lower the temperature is a thumb rule of baking.
- Rich cakes contain more amounts of fat and eggs, and they acquire all its aeration during creaming of fat and sugar. These cakes contain very little baking powder, if at all.
- If rich cakes are baked at high temperature, there will be faster crust formation on cakes. The crust will prevent heat from penetrating inside thus resulting in an under baked product.
- In a lean formula contain fewer amounts of fat and eggs are added. All the aeration in such cakes is achieved by baking powder. Lean cake-batter is thinner than rich batter. Such cakes are baked at higher temperature so that the evolution of gas from baking powder, acquiring of volume by cake and setting of structure of cakes take place simultaneously.
- Baking temperatures: Most cakes are baked at 180 degree centigrade to 200 degree centigrade.
- Batter doesn't contain much moisture and is comparatively stiffer. Such cakes will have a very slow and gradual rise in the oven in order to get thorough baking; and this is precisely the reason for baking them at low temperature.
- If rich cakes are baked at high temperature, there will be faster crust formation on cakes. The crust will prevent heat from penetrating inside thus resulting in an under baked product.
- In a lean formula contain fewer amounts of fat and eggs are added. All the aeration in such cakes is achieved by baking powder. Lean cake-batter is thinner than rich batter. Such cakes are baked at higher temperature so that the evolution of gas from baking powder, acquiring of volume by cake and setting of structure of cakes take place simultaneously.
- Baking temperatures: Most cakes are baked at 180 °C to 200 °C.

Leavening Action in Cakes is achieved by the following methods

- **Mechanical:** During creaming the mixture it is filled with air cells which expand under the action of heat and exert upwards pressure giving volume.
- **Chemical:** Baking powder when moistened with water and heated evolve CO_2 gas during the process of expansion imparts volume.
- **Vapour Pressure:** Water within the mixture forms vapour under pressure a result of which cake is leavened.
- **Biological:** When yeast is used as a leavening agent, produces CO_2 by multiplying in the presence of yeast foods, and water.

For cake manufacture flour of the below mentioned quality needs to be selected: Low protein flour is usually preferred for cake manufacture. This is so because the rise in cake is dependent on the aeration rather than on gluten development. The flour should be finely ground because in batter formation the flour particles do not disintegrate to the same extent as in mixing of dough. In case of high ratio cakes, the sugar increases the gelatinization temperature of starch causing problem in crumb-setting which ultimately cause the cake to collapse immediately after baking. In such case chlorinated flours are preferred. Shortening is responsible for a more tender structure and preventing the dry mouth feel of cakes.

While selecting shortening for cake manufacture the following points should be taken into consideration: Fat should be plastic in nature. It should be solid at room temperature as well as during creaming process. Granular fat should be avoided as it has very poor whipping quality. In order to acquire specific characteristics in cakes, a combination of fats like hydrogenated shortening, butter or margarine may be used. The texture of the fat should neither be too hard nor too soft as very hard shortening will not cream up well, while too soft shortening will not be able to retain aeration.

Table 1 General Cake Formulation

S. No.	Ingredient	% of flour
1.	Flour	100
2.	Sugar	90
3.	Shortening	50
4.	Milk	65
5.	Egg	65
6.	Sodium bicarbonate	1.11
7.	Sodium acid pyrophosphate	1.35
8.	Monocalcium phosphate	0.2

Cakes are leavened either by mechanical aeration of fat and sugar leading to entrapping air cells in the mixture or by using leavening agents. For cake manufacture usually double-acting baking powder is used. Eggs, egg white (ovalbumin), and to a lesser extent milk proteins are important foam stabilizers, which slow down the coalescence of air bubbles. Emulsifiers and egg proteins reduce the foam-destabilizing effect of fat in foam type cakes by keeping the emulsified fat particles well-dissolved in the aqueous phase and preventing from destabilizing the thin foam lamella between the gas bubbles. These also enhance the incorporation of air into the batter during mixing by reducing surface tension.

Broadly, there are two types of cakes - shortened and unshortened. Shortened cakes contain shortening or butter as an essential ingredient which aid in leavening along with baking powder. Examples are chocolate, cake, pound cake etc. Unshortened cakes do not contain fat as a basic ingredient. They are leavened by air or steam. Example is sponge cake.

Types of Methods

There are different methods manufacturing cake:

1. Sugar batter method
2. Flour batter method
3. Blending method
4. Boiled method
5. Sugar water method
6. All in process Sugar batter method

1. Sugar Batter Method

Creaming: In case of using different shortenings, these should be first creamed together in order to blend them thoroughly. Sugar is then added gradually avoiding the adverse effect in the aeration process. Very light texture and brighter appearance of the mixture indicates that the adequate aeration has been achieved.

Addition of dry ingredients: The well mixed blend of flour, baking powder and other dry ingredients is added to the mixture in portions. To avoid toughening of gluten, after addition of dry ingredients there should be minimum possible mixing action. Hence mixing of batter at low speed is essential for retaining the entrapped air as well as toughening of gluten.

Addition of liquid ingredients: When mixing of dry ingredients complete the liquid ingredients are added. Addition of liquids enhances the fluidity of batter to an optimum level required for gradual and uniform leavening of cake during baking operation.

Baking: The batter is poured on baking pan and baked at 15-180 °C for 30 minutes. The baked cake is then cooled and packaged/consumed.

2. Flour-Batter Method

The flour batter method is specially suited for making lean cakes which do not contain much fat or egg and most of aeration is achieved through baking powder.

Creaming of fat and flour: Fat and equal quantity of flour is creamed together till a light and fluffy mixture. During creaming the flour is added gradually. The advantage of this step is that as a major portion of the flour is coated with fat before addition of any liquid, the gluten development is avoided when the flour is mixed with liquid. Due to the same reason slightly strong flour may also find use in cake manufacture.

Addition of whipped mixture of egg and sugar: Egg and an equal quantity of sugar are whipped to a stiff froth. This is added in small portions to the creamed mixture of fat and flour to avoid the curdling of batter and after each addition, mixed thoroughly.

Addition of liquids: The remaining sugar is dissolved in milk or water and added to the mixture. Any colour or flavour is also added along with this liquid.

Addition of remaining flour: The properly mixed blend of remaining flour and baking powder is added and mixed gently. At this stage vigorous mixing may induce knocking out of air cells resulting into poor cake volume.

3. Blending Method

The blending method is used for manufacture of high sugar to flour ratio cakes. The process involves the following steps

- Whipping of shortening, flour, baking powder and salt
- Addition of mixture of sugar, liquid, colour and flavour
- Addition of eggs and mixing
- Pouring of batter into baking pan
- Baking
- Cooling

4. Boiled Method

The process involves the following steps

- Melting of shortening in a bowl

- Addition of a portion of flour
- Addition of properly whisked egg-sugar sponge in parts
- Addition of remaining flour
- Pouring of batter into baking pan
- Baking

5. Sugar-Water Method

The cake from sugar-water has more aeration and better emulsification resulting into better texture and longer shelf life. The process involves the following steps:

- Dissolving sugar in half of the quantity of water
- Addition of remaining ingredients (except egg) and agitation for aeration
- Addition of egg and mixing
- Pouring of batter into baking pan
- Baking
- Cooling

6. All in Process

- Mixing all the ingredients together
- Aeration of mixture with control on mixer speed and time
- Pouring of batter into baking pan
- Baking

After adding all the ingredients, the mixing is controlled as follows:

Half a minute at low speed: This is done so that all the dry ingredients are moistened without flying off from the bowl.

Two minute at fast speed: All the ingredients break and are incorporated evenly hroughout the mass. The batter is also well aerated.

Two minute at medium speed: Aeration achieved during the second stage is not evenly disturbed in the batter. By mixing at medium speed the large air cells break up into smaller cells and the aeration of the mixture becomes even.

One minute at low speed: This is done in order to eliminate any possible large air pockets and still finer breaking down of air cells.

Types of Cakes

All mixing methods can be divided into two categories: high fat those that create a structure that relies primarily on creamed fat and egg foam- those that create a structure that relies primarily on whipped eggs. Within these broad categories are several mixing methods or types of cakes.

Whipped Cakes

Cakes based on whipped egg foams include European-style genoise as well as sponge cakes, angel food cakes and chiffon cakes. Some formulae contain chemical leaveners, but the air whipped into the eggs is the primary leavening agent.

Genoise

This is the classic European-style cake. It is based on whole eggs whipped with sugar until very light and fluffy. Chemical leaveners are not used. A small amount of oil or melted butter is sometimes added for flavour and moisture. It is often baked in thin sheets and layered with butter cream, pureed fruit, jam or chocolate filling to create multi layered specialty. It is usually soaked in flavoured sugar syrup as the cake is dry.

Genoise Cake

Pre heat the oven and prepare the pans.

- Sift the flour with any additional dry ingredients.
- Combine the whole eggs and sugar in a large bowl, warm over a double boiler to a temperature of 38 degree centigrade.
- Whip the egg and sugar mixture until very light and fluffy.
- Fold the sifted flour into the whipped eggs carefully.
- Fold in oil or melted butter.
- Divide the batter immediately into pans and bake at 180 degree centigrade approximately for 8 minutes.

Angel Cake

Angel food cakes are tall, light cakes made without fat and egg yolk but leavened with a large quantity of whipped egg whites. They are traditionally baked in un-greased pans but large loaf pans can also be used. The pans are left un-greased so that the batter can cling to the side as it rises. The cake should be inverted as soon as they are removed from the oven and left in the pan to cool. This technique allows gravity to keep the cakes from collapsing or sinking as they cool. They contain no fat, but angel food cakes are not low in calories as they contain high % of sugar. The classic angel food cake is pure white but flavourings, ground nuts or cocoa powder, may be added. They are rarely frosted. They may be topped with fruit flavoured or chocolate glaze. They are often served with fresh fruit or whipped cream.

Method

- Preheat the oven.
- Sift the dry ingredients together.
- Whip the egg whites with sugar (half the quantity of sugar) until foaming, add cream of tartar and beat to soft peaks.
- Gently fold dry ingredients into the egg whites along with remaining sugar. Spoon the batter into an un-greased tray at 180 °C for 35–40 minutes. \
- The cake surface will have deep cracks.
- Remove the cake from oven and immediately invert the pan. Rest till it cools.

Whisked Cakes

They are also known as biscuit whisked cakes or sponge cakes. They are made with whole separated eggs. Batter is prepared with egg yolks and ingredients, then the egg whites are whipped to form peaks with a portion of sugar and folded into the batter. They are primarily leavened with air, but baking powder may be included in the formula. Sponge cakes are extremely versatile- they can be soaked with sugar syrup or liqueur and assembled with butter cream or whipped cream as a traditional layered cake. They can also be layered with jam, custard, cream fillings.

Sponge Cake

Method

- Line two greased aluminium baking trays with brown paper.
- Sift the flour and set aside.
- Separate the eggs, placing the yolks and whites in separate mixing bowls. Whip yolks on high speed for 3-5 minutes, until thick, pale and at least double in volume. Whip in the vanilla essence and cream of tartar. The yolks should be whipped to ribbon, that is until they fall from the beater in thick ribbons that slowly disappear into batter surface.
- Beat the egg whites till fluffy add cream of tartar and 2 tbsp of sugar. Peak the egg white.
- Pour the egg yolk into whites, quick fold the both mixtures, sprinkle the remaining sugar over the mixture and fold in.
- Fold in 1/3 of sifted flour. Repeat the procedure until all of the flour is incorporated. Do not over mix; fold just until incorporated.
- Pour the batter in prepared pans. Bake immediately at 190 °C for 30 minutes.

Chiffon Cake

Although chiffon cakes are similar to angel food cakes in appearance and texture, the addition of egg yolks and vegetable oil makes them moister and richer. Chiffon cakes are usually leavened with whipped egg whites but may contain baking powder as well. Like angel food cake, chiffon cakes are baked in an ungreased pan to allow the batter to cling to the pan as it rises. Chiffon cakes can be frosted with a light butter cream or whipped cream or topped with a glaze. Lemon and orange chiffon cakes are the most popular, but formulae containing chocolate, nuts or other flavourings are also common.

Method

- Sift together the flour, half the sugar, baking powder and salt.
- In a separate bowl mix the oil, yolks, water, juice, zest and vanilla. Add the liquid mixture to the dry ingredients.
- In a clean bowl beat the egg whites until foamy. Slowly beat in the remaining 170 gm of sugar. Continue beating until the egg whites are stiff but not dry.
- Stir one third of the egg whites into the batter to lighten it. Fold in the remaining egg whites.
- Pour the batter into a greased 10 inch tube pan. Bake at 160 °C until a toothpick comes out clean for approximately one hour.
- Immediately invert the pan over the neck of a wine bottle. Allow the cake to hang upside down until completely cool and then remove from the pan.

Butter cake

Pound cake

Sponge cake

Genoise cake

Angel Food cake

Chiffon cake

Cup Cakes

Chocolate satin cake

Fig 5.1 Various types of cakes

Creamed Fat Cakes

Creamed-fat cakes include most of the popular American-style cakes: pound cakes, layered cakes, coffee cakes and even brownies. All are based on high-fat formulas containing chemical leavening agents. A good high-fat cake has a fine grain cell of uniform size and a crumb that is moist rather than crumbly. Crusts should be thin and tender.

Creamed-fat cakes can be divided into two classes: butter cakes and high ratio cakes.

Butter Cakes

Butter cakes, also known as creaming method cakes, begin with softened butter or shortening creamed to incorporate air cells. Because of their high fat content, these cakes usually need the assistance of a chemical leavening agent to achieve the proper rise. The classic American layer cakes, popular for birthdays and special occasions are made with the creaming method. These cakes are tender yet sturdy enough to handle rich butter creams or fillings. High-fat cakes are too soft and delicate, however, to use for roll cakes or to slice into extremely thin layers. When making butter cakes, the fat should be creamed at low to moderate speeds to prevent raising its temperature. An increased temperature could cause a loss of air cells.

Method

- Sift the cake flour, baking powder and salt together
- Cream the butter and sugar until light and fluffy. Add the eggs one at a time, beating well after each addition. Stir in the extract.
- Fold in the dry ingredients. Divide the batter into greased loaf pans.
- Bake at 160 °C until golden brown approximately for 1 hour 10 minutes.

High Ratio Cakes

- High ratio cakes received its name from the usage of ingredients. It contains a high ratio of sugar and liquid to flour and therefore these cakes are known as high ratio cakes. They have a very fine, moist crumb and relatively high rise. High ratio-cakes are almost indistinguishable from modern butter cakes and may be used interchangeably.

Method

- Combine the flour, sugar, shortening, salt, baking powdered milk, corn syrup and 1 litre cold water in a large bowl of a mixer fitted with a paddle attachment. Beat for 5 minutes on low speed.
- Combine the remaining ingredients in a separate bowl. Add these liquid ingredients to the creamed fat mixture in three additions. Scrape down the sides of the bowl after each addition.
- Beat for 2 minutes on low speed.
- Divide the batter into greased and floured pans. Pans should be filled only half way. One gallon of batter is sufficient for an 18 inch × 24 inch × 2 inch sheet pan. Bake at 170 °C until a cake tester comes out clean and the cake springs back when lightly touched, approximately 12–18 minutes.

Cheesecake

Types of cake cannot be completed without mentioning cheese cake. Cheesecakes, which are almost as old as western civilization, have undergone many changes and variations since the ancient Greeks devised the first known recipe. Americans revolutionized the dessert with the development of cream cheese in 1872.

Cheesecake is baked custard that contains a smooth cheese, usually a soft, fresh cheese such as cream, ricotta, cottage or farmer cheese. A cheese cake may be prepared without a crust or it may have a base or sides of short dough, cookie crumbs, ground nuts or sponge cake. The filling can be dense and rich or light and fluffy. Fruit, nuts and flavourings may also be included in the filling. Cheese cakes are often topped with fruit or sour cream glaze.

High Ratio Cakes

High ratio cakes are baked at low temperature. They achieve all its aeration during creaming of fat and sugar or during whipping of eggs. They contain very little baking powder. The batter does not contain much moisture and is comparatively stiffer. Such cakes should have a very slow and gradual rise in the oven in order to get thorough baking.

If baked at high temperature there will be faster crust formation on cakes.

The crust will prevent heat from penetrating inside the cake resulting in an under baked product with a dark colour crust. The crust may burst open spoiling the appearance of the cake. Good results are achieved by placing a vessel of water in the oven. The water consumes some of the heat and at the same time the water delays the process of crust formation on cakes thus facilitating even volume.

In this the quantity of sugar is more than flour. Emulsified type of shortening and special cake flours are used for making high ratio cakes. Aeration is achieved by creaming fat.

Lean Cakes

They do not contain much fat or egg and much of aeration is achieved by using baking powder. So after addition of flour vigorous movement can be given. Lean cake batter is thinner than rich cake batter and such cakes are baked at high temperature. This ensures that evolution of gas from baking powder, acquiring volume by cake and setting of structure of cakes takes place. If such cakes are baked at low temperature, there will be evolution of CO_2 but due to low temperature the structure of cake will not set and the cake will collapse.

Common Faults

Reasons for faults in cakes may generally be grouped as follows:

- Wrong quality of raw material
- Improper balance of formula
- Operational mistakes

Wrong Quality of Raw Material

1. Flour

- Usage of strong flour for cake making results in gluten formation which results in small volume, peaked top and unsightly crack in the centre and an uneven texture.
- Strong flour will lead to shorter shelf life.

- Too weak flour will not be able to carry sugar and fat and the cakes will be poor in volume.

2. Sugar

- Very large crystals of sugar will not dissolve during mixing and the resultant cake will have all the defects as if lack of sugar has been used in the formula. i.e., harsh crumb.
- It leads to poor consuming quality and staling.
- It is a must that sugar and water should be proportionate other wise it leads to the above mentioned faults.
- Too large or too small crystals of sugar are not desirable as they do not cream up well i.e., poor aeration and thus adversely affect the volume and texture of cakes.

3. Fat

- Granular fats will not cream up well and are not capable of holding the air cells. Resulting cakes will be poor in volume and have a coarse texture.
- If the fat melts during mixing operations, aeration will be lost, affecting the volume and texture adversely.

4. Eggs

- Weak and watery eggs or staled eggs should never be used as it will not have good whipping quality.
- Weak and watery eggs will result in curdling of batter.
- Curdling of batter will lead the aeration to escape and will result in poor volume and texture.

5. Baking Powder

- Humid or wet baking powder will not achieve the desirable aeration that results in improper volume of cake.

Improper Balance of Formula

1. Sugar

- Excess usage of sugar will lead to collapse in the centre.
- It will result in dark, hard upper crust.
- Harsh crumb will result due to lack of moisture.

2. Fats

- Excess of fat will result in improper volume.
- Lack of fat will result in dryness, tunnels and will stale rapidly.

- Operational Mistake:
- Improper sifting of dry ingredients will lead to substandard shape and texture.
- Lack of aeration will result in poor volume.
- Insufficient moisture content in batter will lead to development of gluten and result in improper texture and shape.
- Care should be taken while folding flour into the creamed mixture to prevent escaping of aeration.

References

Edmund, B. James Stewart and Barr G.S.T. Cake Making. Leonard Hill Books, London 1966.
FAO Repository, Cakes and Biscuits, FAO, United States of America F.

6

Formulations and Processing of Breads

All bread is made by baking a dough that has two basic ingredients, flour or meal and a liquid. Bakers can use a wide variety of both components. The most common type of flour used for bread and most other baked goods is made from wheat. Wheat flour has a pleasant taste and contains a large amount of an elastic protein substance called gluten. Gluten aids in baking uniformly light bread that rises (swells) properly. Other baking flours are made from barley, rye, corn, rice, oats, soybeans, and potatoes. These flours, particularly soybean flour, may equal wheat nutritionally, but none can match wheat for creating light, even-textured bread. Hard wheat flour makes lighter bread than does soft wheat flour because it is richer in gluten. Rye and whole wheat breads are made lighter by adding white flour. The liquids used in baking include water, sweet or sour milk, yogurt, wine, and beer.

Bread is either leavened or unleavened. Leavened breads contain some substance that produces bubbles of carbon-dioxide gas. These gas bubbles inflate the dough, causing it to rise and become light and porous. Most kinds of basic breads are leavened with a fungus called yeast. Biscuits, muffins, and cakes and other pastries are leavened with either baking powder or baking soda. Unleavened bread is dry and hard. Familiar kinds of unleavened breads include water crackers, the rye crisp of Sweden, and Jewish matzoth. Whether leavened or unleavened, most breads contain other ingredients in addition to flour and a liquid. An almost limitless variety of breads can be made by adding a sweetener, shortening, cheese, eggs, meat, fruit, vegetables, seeds, or nuts. A sweetener, either sugar or syrup, is used in almost all bread for its taste or as an aid to yeast growth. Bread may also have an external sweetener in the form of a decorative glaze. The high fat content of shortening and cheese increase tenderness and flakiness in bread. Perhaps the best example is the French croissant. Eggs help leaven bread dough by adding to the bread's lightness. They can be brushed on top of the dough before baking to create a shiny crust, as in the Jewish hallah. Some breads from many nations contain fruit, a vegetable, meat, seeds, or nuts. Examples include the fruit scone of Great Britain, the spinach paratha of India, the Southern sausage bread of the United States, the Easter sesame bread of Greece, and the almond sweet bread of Finland.

Essential Ingedients used in Bread making

Flour

The flour is the main ingredient used in bread making. Strong flour is recommended for bread making and it should have a creamy white colour, it should feel slightly coarse when rubbed between the fingers. The protein content of the flour should be high. Flour provides the structure to bread. Flour contains proteins that interact with each other when mixed with water, forming gluten. It is, this elastic gluten framework which stretches to contain the expanding leavening gases during rising. The protein content of a flour affects the strength of a dough. The different wheat flour types contain varying amounts of the gluten forming proteins. In yeast breads, a strong gluten framework is desirable.

Types of flours used in Bread making

Wheat flour is consumed in larger quantities worldwide than any other cereal flour. This is because of its extensive availability wheat can be grown under widely varying climatic conditions and to its almost universal acceptance as a staple food item (see Wheat). Wheat flour contains a unique protein called gluten. When wheat flour is mixed with water, the gluten forms elastic dough. When the dough is baked in a hot oven, it expands to several times its original volume. Flours made from soft wheats containing less than 12 percent of gluten protein are used to make tender products such as cakes and crackers. Flours made from hard wheats containing more than 12 % protein are used for bread and roll production. The miller can supply the baker with a wide range of wheat flour types, each custom milled to the baker's specifications

Rye flour contains a small amount of gluten protein and may be used by itself to produce dark rye breads. It is often blended with wheat flour to produce finer textured, light rye breads. The distinctive flavor of rye flour makes it a common inclusion in such items as snack foods and prepackaged toast. Rye has always been grown amid wheat, being a more heartier and insistent plant-like a weed. At one time the European farmers tried to remove it from cultivated wheat crops but they had to give up, because the rye persisted growing at a faster rate than they could remove it. So they simply started harvesting it with the wheat, and started calling it rye flour. Nowadays rye is grown separately and one can get 100 percent rye without the adulterating wheat. Rye contains very little gluten so rye breads are often created by a high proportion of gluten strong wheat flour to assure a well-risen loaf. Breads that are completely rye will not rise much, but they will have a real hearty intriguing flavour. Rye berries are dark brown whole grains. Although each grain delivers about 7 percent protein but virtually no gluten, the fragrant benefit of adding rye to breads is remarkable. One can even add the soaked softened whole grain and the cracked rye berries to breads for extra nutrition, distinct toasted aroma, and subtle bite in flavour. Some mills grind the whole rye berry, bran and all into a very coarse meal, which is often called pumpernickel rye flour. (Pumpernickel bread does not come from the pumpernickel grain. The name derives from a German, Herr Pumpernickel, who popularized this dark, hearty bread many years ago.)

Corn flour and **Corn meal** are used in the production of crusty corn breads and muffins. Corn has no gluten but does have a distinctive flavor and a pleasant yellow color that is desirable in many products.

CORN is native to the Americas. According to the archeologists the Aztecs grew sacred forests of these yellow grain. The kernel just like the wheat berry is full of complex carbohydrates. Cornmeal comes in a variety of different colours. Mostly one finds the yellow cornmeal, though one even gets the white variety as well as the blue cornmeal. The yellow and the blue are the most flavourful, but you may have to add a little more fat to the blue variety. Cornmeal adds a rich sweet taste to the crumb and gives a rugged look to the crust.

Oat flour and **Oat meal** are used primarily in breakfast foods and granola-type products. Oat flour is the most nutritionally complete of all flours. Oats the grain is called Groat, and it is packed with B vitamins, vitamin E, minerals, and iron, a great deal of soluble fiber and a fair amount of fiber. One can use the familiar rolled oats, which is often cooked into oatmeal, on top of some bread for the classic country appearance.

Barley flour can be found in baby foods and malted milks. In some countries large quantities of barley flour are used for bread making.

Sorghum and millet flours are popular in India, Central America, and Ethiopia. They are utilized in the making of flat bread, tortillas, and pancakes.

Rice has long been the staple food of Asia. It is normally eaten as a whole grain, so rice mills remove either the hull to produce brown rice or both the hull and the bran coat for the production of white rice. A small percentage of rice is converted to flour and is used in baby foods and sauces.

Buck wheat is not a true cereal grain, but buckwheat flour provides a distinctively flavored pancake and breakfast food ingredient. Buck wheat is actually a grass like herb, related to sorrel, and has a slightly sour flavour. Buckwheat flour has an attractive tan colour speckled with dark brown.

Whole Wheat Flours are made from 100 percent whole-wheat berries creating heady light brown flour. The flour feels gritty when rubbed between your fingers, and the bran is visible in flecks. During fermentation these bran flecks help the dough rise as they trap the carbon dioxide.

Bleached and Unbleached Flour: Unbleached flour is ground wheat flour with the bran sifted out, but with the entire germ remaining. As the sifted flour is stored and aged it naturally starts bleaching and lightens in colour. It was found that the aged bleached flour was easier to manipulate in the doughs. Commercial bakeries began to demand the whitened flour. To accommodate this need without the lengthy aging period, chemical methods of bleaching were developed using bromates, iodates and chlorine dioxide etc.,

Semolina Flour is granular flour made from Durum, which is a variety of hard wheat, with the bran and germ removed. The term Semolina refers to the texture, which is similar to finely ground cornmeal. It is believed that semolina flour makes the best pasta. Semolina Flour is used in making pasta and Italian puddings. It is made from durum wheat, the hardest type of wheat grown. The flour has highest gluten. When other grains, such as rice or corn, are similarly ground, they are referred to as "Semolina" with the grain's name added, i.e., "Corn Semolina" or "Rice Semolina." There are different grades:

1. Semolina flour is finely ground endosperm of durum wheat.
2. Semolina meal is a coarsely ground cereal like farina.
3. Wheatena is ground whole-grain wheat.
4. Durum flour is finely ground semolina and is grown almost exclusively in North Dakota.

Organic Flours are flours that are grown in fields that are fertilized by naturally occurring substances-composted materials, aged animal manure, and green manure. The soil as a result becomes more loose and this allows the roots to grow deeper, and the plant structure is strong. When the grain is harvested and cleaned it is stored without the use of fumigants or synthetic agents. The result is a full bodied wholesome taste, like a tomato or a carrot grown in your own garden

Yeast is next important ingredient to flour for bread making. In olden days most of the bakers used barn method of bread making in which wild yeast was cultured. It was necessary to use prolonged fermentation. Due to this, bread had that peculiar fermented flavour which is still remembered by people nostalgically. However times have changed, bakers yeast is easily available and baker's botherations about uncertainties of fermentation have been eliminated.

For practicing bakers, it is necessary to understand the functioning of yeast, so that baker is in a better position to control yeast activity in doughs and thus control the quality of bread and other fermented products. Bread doughs are fermented basically for two reasons i. e

1. Production of carbon dioxide gas which gives volume to the product.
2. For maturing or conditioning the dough (gluten) so that it attains sufficient mellowness to stretch under the pressure of carbon dioxide gas and form the structure of the product.

These functions are performed by yeast. Yeast is a unicellular microscopic plant. Its structure consists of a cell wall, protoplasm, and vacuole. It requires food, moisture and temperature for its growth and reproduction. Yeast multiples by budding. When yeast cell is placed in a liquid medium at optimum temperature containing simple sugar, then the cells starts growing buds on its cell wall which keeps on growing until daughter cells acquire the same size as mother cell and start producing other buds. Yeast is a living micro organism until it is destroyed by heat. The protoplasm of yeast contains certain enzymes like Inverase, Maltase, zymase and protease by which fermentation activity of yeast is made possible. Hence enzymes are known as catalytic agents. There is no other organism which contains the same combination of enzymes in the same proportion. That is why there is no substitute for yeast as a fermenting agent.

Composition of yeast is as follows:

Proteins	–	14.00 %
Carbohydrates	–	10.20 %
Fats	–	0.46 %
Minerals	–	2.34 %
Moisture	–	73.00 %
Enzymes		Present
Vitamins		Present

Yeast is available in 3 forms

1. Fresh yeast / Compressed / wet yeast is moist and perishable.
2. Active dry yeast is a dry granular form of yeast. It has to be activated before use, i.e., it has to be rehydrated in 4 times water its weight of warm water before use.

3. Instant dry yeast is also a dry granular form of yeast, but it does not have to be dissolved in water before use. It can be added in its dry form because it absorbs water much more quickly than regular dry yeast. Compressed yeast should be used 2–2.5 times more as compared to dry yeast.

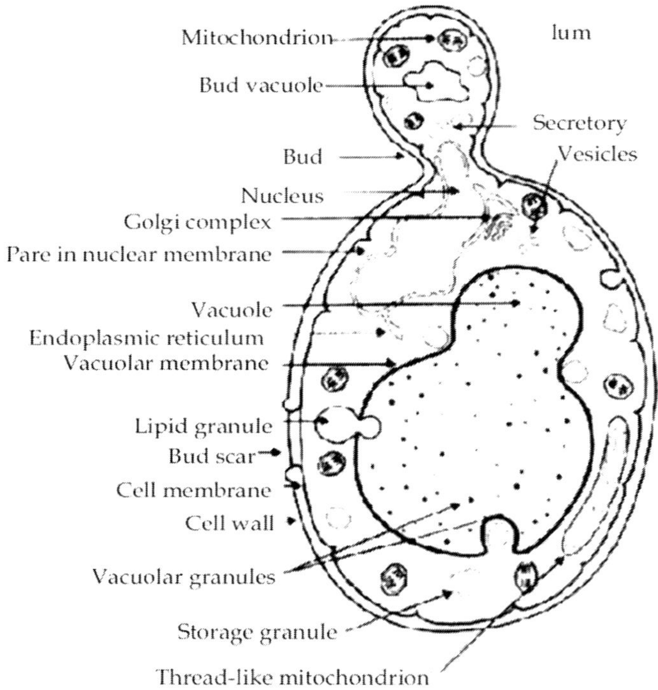

Fig 6.1 Structure of yeast

Water

Water binds together the insoluble Proteins of flour, which form gluten. Any water which is fit to drink can be used for bread making. However; it should be remembered that water, being a very good solvent, is rarely found in nature without any minerals dissolved in it. Hard water contains more minerals than soft water. These minerals in limited quantities, have a beneficial effect on gas production as the yeast requires minerals for vigorous fermentation. The gas retention of the dough is also improved as minerals have a tightening action on gluten. For this reason precisely, very hard water should not be used for bread making. With very hard water, it may be necessary to increase the quantity of yeast and reduce salt to appropriate level. Lactic acid could be used which will have a mellowing effect on gluten. The quantity of lactic acid should be carefully regulated because it has a very pungent odour which may interfere with the pleasant flavour of bread. Very soft water is also not desirable as it is not conductive either to good gas production of gas retention and in such cases it may be necessary to increase the content of mineral yeast food (MYF). Medium hard water (Hardness about 17ppm) is considered to be most suitable for bread making.

Salt

Salt imparts taste to bread. It is also one of the most important constituent to bring out the flavour in bread. It has a controlling effect on yeast activity and thus keeps the fermentation speed under check. As the salt has tightening action on flour proteins, it improves the gas retention power of the dough. Being hygroscopic substance, it helps to keep the bread fresh and moist for longer time. The colour of crust is largely dependent on the amount of sugar present in bread at the time of baking, and this amount of sugar will depend on the speed of yeast activity, which , in turn is controlled by salt. Therefore, if there is less salt in formula, yeast action will be more than normal and there will be less sugar left for caramelization resulting in poor crust colour. Conversely, more salt in formula will produce a bread with harsh red crust colour, as there will be more sugar left at the time of baking due to check on yeast activity. The quantity of salt in bread formula will vary between 1.25 to 2.5% depending on the strength of flour, length of fermentation time, hardness of formula water, the level of flavour desired in the product and the constitution of MYF. In sweet fermented products, the salt content generally varies from 1.0 to 1.5%. If salt is omitted or reduced, other spices or flavorings in the recipe should be increased slightly. In yeast dough, salt slows yeast fermentation. Omitting or reducing the amount of salt in yeast dough can cause the dough to rise too quickly, adversely affecting the shape and flavour of bread.

Sugar

The main function of sugar in bread making is to provide for yeast which in turn produces carbon dioxide gas, that raises the dough fabric. It also helps in enhancing the flavor of bread. Being hygroscopic substance, sugar helps to retain moisture in bread. It contributes to the golden brown outer crust colour of bread. Apart from the sugar added in the formula, there is also another source of sugar in fermenting dough and that is by the activity of diatase enzyme on starch. The capacity of flour to produce sugar from starch is known as diastatic capacity and is carried out by a group of enzymes known as diastase which are contained in flour. Diastase enzyme converts starch into maltose sugar and dextrins. Maltose is broken down by another enzyme maltase into dextrose (glucose) which provides food for yeast at the critical time of final stage of fermentation i.e during initial stage of baking. It also imparts bloom to the bread. At times, flour may be deficient in diastatic activity and bread may not get oven spring. To correct such condition one of the remedies may be tried.

1. Make a small quantity overnight sponge and add to the dough. This will reduce the pH of the dough.
2. Use diastase malt at the rate of 0.225%.
3. Blend with flour having good diastatic activity.

Fats and Oils

Fat is used in bread doughs at the rate of 1–2%. To that extent it improves the nutritional value of bread. In small quantities it has a lubricating effect on gluten strands, thus improving their extensibility which enables the bread to acquire good volume. In larger quantities (more than 6%) fat exerts a dead weight on fine web like structure, thus hampering the volume of bread. Fat also helps in retention of moisture in bread and improves its slice ability. Fat should be added during the last stages of mixing. If it is added in the beginning, it will have an adverse effect on water absorption power of the flour. Fats can be used in the form of solid shortening, margarine, or butter; or in the liquid form of oil contributes tenderness, moistness, and a smooth mouth feel to the bread. Fats enhance the flavors of other ingredients as well as contributing its own flavor, as in the case of butter. In baked goods such as muffins, reducing the amount of fat in a recipe results in a tougher product because gluten develops more freely. Another tenderizing agent such as sugar can be added or increased to tenderize in place of the fat. A small amount of fat in yeast dough helps the gluten to stretch, yielding a loaf with greater volume.

Eggs

Eggs serve many functions in bread making. They add flavor and colour to it. Egg contains proteins, fat and lecithin which helps in keeping the bread moist and soft due to their modifying action on gluten. Protein of egg has strengthening action on flour proteins which improves volume and crumb structure of bread. As an improver, egg can be used at the rate of 4 to 6% based on flour. Under the action of heat, egg white coagulates faster than egg yolk and beneficial components like fat, lecithin are found in egg yolk only. Hence it is advisable to use only egg yolk in fermented goods. However, egg as an improver in bread should be used with caution as it may be acceptable to majority of bread consumers.

Bread Improvers

If the quality of raw material is good and the baker knows his job well, it is hardly necessary to use any bread improvers. However, ideal conditions for bread production do not always exist. Quality of flour varies from consignment to consignment, mineral content of water varies from place to place and with the complete mechanization of bread production process, it has become unavoidable to make use of certain chemicals in order to ensure consistently good quality of the product. Flours, always do not contain desirable quality gluten forming proteins. Any added material which can improve the strength and extensibility of gluten of flour is known as Bread improver.

Potassium Bromate is one of the earliest known chemical used by baking industry to improve the quality of bread. It oxidises the the gluten, giving it more strength, which has direct bearing on oven spring and other related characteristics of bread. After reaction is complete potassium bromide remains as end product which is considered to be harmless substance. The quantity of potassium bromate to be used ranges from 10 to 40 ppm (parts per million), depending on the strength of maida, formulation of the product and process used.

For convenience, a definite quantity of potassium bromate can be dissolved in measured quantity of water and the solution can be used to provide potassium bromate at desired level.

For e.g., Potassium bromate 10 g + 1000 mL of water

If 100 ml of this solution is used in 100 kg maida, the quantity of potassium bromate will be 1 g/100 kg which will amount to 10 ppm. Potassium bromate can also be diluted with maida in desired proportion as given below:

For eg., Potassium bromate 10 g + 4990 g flour

From Improved Flour 5000 g, 500 g is to be added to 100 kg and the formula will provide potassium bromate at the level of 10 ppm.

Milk

Milk has tightening action on flour proteins which eventually improves the texture of bread. Inclusion of milk in the formula necessitates addition of extra 2–3% water in the dough thereby increasing the yield by that amount. Milk improves the flavour and taste of bread and lactose content of milk improves the crust colour. In regular white bread 1–2% milk(solids) can be added for improving the quality of bread which will also subscribe to its nutritional content. However, to qualify as milk bread, it should have higher amounts of milk solids content. Milk can be used for bread making in any form i.e. fresh liquid, evaporated, condensed or powder form. Skimmed milk is preferred because due to absence of fat it has better shelf life and it is less expensive than whole milk powder.

Bread Making Methods

The Baking Process

The basic process of baking yeast bread starts with measuring and mixing the various ingredients to make the dough and adding yeast so that it rises. The dough is then kneaded to develop the gluten and is again allowed to rise. The kneading and rising steps may be repeated several times. Next, the dough is shaped into a loaf and baked. Baking cooks the dough, firms the loaf and forms a crust on it, and improves the flavor. Finally, the loaf of bread may be sliced before being wrapped. Commercial bakeries have machines that do the work of measuring, mixing, kneading, baking, slicing, and wrapping. Skilled bakers run the machines, and nothing is left to chance. The ingredients are weighed precisely, the temperature and humidity are closely monitored, and the individual steps of the baking process are carefully timed. Every bakery uses a special blend of flour, produced by mixing the wheat before or after it has been milled. In most large bakeries the manufacturing process begins in bins on a high floor so that gravity can draw the flour or dough from one machine down to the next. After a final sifting, the flour is fed into a scale that automatically weighs the right amount and pours it into a mixer on the floor below. Water or liquid fraction is poured in to form dough, and yeast and other ingredients are added. The amount of flour used to make the dough can be affected by the temperature and humidity in the bakery. In addition, the temperature of the water must be exactly correct to dissolve the yeast. The yeast will be killed if the water is even slightly overheated. On the other hand its growth will be stunted by water that is too cold. In the next step of the manufacturing process, the dough flows into huge troughs that are taken into a fermentation room. It is left there to rise for a set amount of time, usually several hours.

Next, a divider scales the dough into pieces of just the right weight for the baking pans. The rounder shapes the pieces into balls, which then move through the overhead proofer. There the dough rests for a few minutes to recover from the rough dividing and rounding processes, thus ensuring tender loaves. The balls of dough drop from the overhead proofer into a molder, which shapes them to fit the baking pans exactly. The filled pans are placed in the proof box, where the final rising takes place. The proof box has a slightly warmer and moister atmosphere than that of the fermentation room. The pans then go into an oven, where they are baked at a temperature of more than 400 F (204 °C) for about 30 minutes. Low-pressure steam is injected into the oven to prevent the crust from forming too quickly. Most large bakeries use reel ovens or traveling ovens. A reel oven looks like an enclosed Ferris wheel, with the pans of bread on rolling racks. In a traveling oven the pans move slowly on a conveyor belt through a long baking chamber, and the bread comes out the other end. Some traveling ovens measure more than 100 feet (30 meters) in length, and they can bake more than 5,000 loaves of bread per hour.

After the loaves have been slowly cooled, a slicer cuts them into uniform slices. Finally, a wrapping machine places moisture-proof paper around each loaf and seals the paper to keep the bread fresh and protect its flavor. The loaves are then packed into trucks and taken to stores. The process of making unleavened bread, which is sometimes called no-yeast bread or quick bread, is much simpler than that used for yeast bread. Since the dough contains no yeast, kneading and rising are not involved. The procedure consists merely of measuring and mixing the ingredients and then shaping the dough and baking it. Bakeries make many products in addition to bread, including rolls, crackers, biscuits, and such pastries as cookies, cakes, pies, and doughnuts. Machines do much of the work in baking these products, as in making bread. Bakers use a variety of devices for molding and cutting and for such operations as making and applying frosting and icing. There are two general kinds of cakes butter cakes and sponge cakes. Butter cakes contain butter or some other fat, plus flour, sugar, eggs, leavening, milk, salt, and flavoring. Bakers make many varieties of these cakes by adding chocolate, molasses, spices, nuts, coconut, or other ingredients. Sponge cakes, such as angel food cakes and similar products, have no fat. They usually consist of flour, eggs, sugar, salt, and flavoring. The eggs provide the liquid, and the air for rising as well, and cream of tartar is added for lightness and tenderness.

I. Straight Dough Method

In this method all the ingredients are mixed together, and the dough is fermented for a predetermined time. The fermentation time of straight dough depends on the strength of flour. Strong flours require more fermentation time to mature adequately. Flours which require 2–3 hours for maturing should be used for making bread by straight method. Flours that take very long period for maturing should not be used in straight method because during prolonged fermentation periods it is very difficult to control the temperature of dough and rise in temperature will invariably cause acid taste and flavour in bread. As temperature rise has immediate effect on fermentation speed, it is very necessary to control the temperature of a straight dough by;

1. Using shorter fermentation periods
2. Adjusting the temperature of doughing water
3. By fermenting the dough at optimum (room) temperature i.e. between 78 deg. to 80 °F

When it is desired to ferment straight dough for longer period, it should be remembered that gluten will soften up to a greater extent and is likely to become sticky, therefore the dough should be made tighter. Yeast content should be reduced but sugar content should be increased in order to provide food during prolonged fermentation. Salt content is increased as it provides stability to the dough and keeps the fermentation speed under control which is necessary during long fermentation period.

Fig 6.1 Direct mixing of ingredients

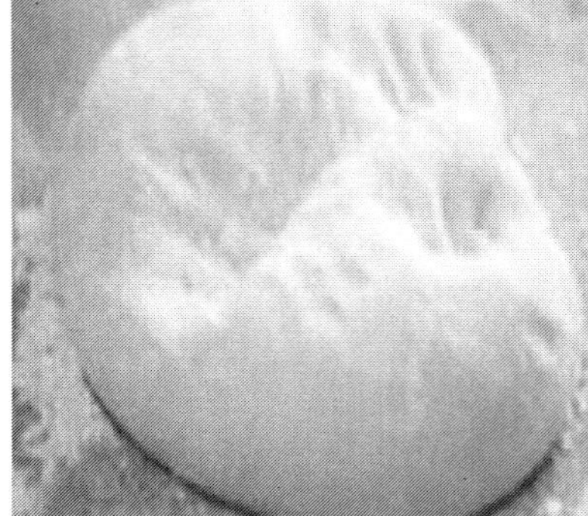

Fig 6.2 Dough after knocking

(A) Salt–Delayed Method

This is a slight variation of straight method, where all the ingredients are mixed except salt and fat. As the salt has a controlling effect on enzymatic action of yeast, the speed of fermentation of a salt less dough will be faster, and a reduction in total fermentation time could be affected. The salt is added at the knock-back stage. The method of adding salt at the later stage may be according to the convenience of individual baker. It may be sifted (dry) on the dough and mixed. It may be creamed with fat and mixed. Whatever way is chosen for mixing the salt, only three-fourth (of the actual mixing time) mixing should be given initially and one- fourth mixing at the time of adding salt. The method is specially suitable if strong flours are to be used for bread making by straight method. Due to absence of salt, the fermentation speed enhanced and gluten is matured in a reasonably shorter time.

(B) No Time Dough Method

In this method, dough is not fermented in a usual manner. It is just allowed a brief period (about 30 min). for it to recover from the strains of mixing. Since dough is not fermented the twin functions of fermentation (i.e. production of gas and conditioning of gluten) are achieved to some extent by increasing the quantity of yeast (2–3 times of original quantity) and by making the dough little slacker and warmer. Although it is possible to make fairly acceptable bread (during emergency) by using this method the product has poor keeping quality and lacks in aroma. Due to absence of fermentation the gluten and starch are not conditioned sufficiently to retain the moisture and there is no flavour because flavour producing bi-products of fermentation are absent. As there is increased quantity of yeast present, the bread may have a strong yeast flavour.

II. Sponge and Dough Method

Previously, in this chapter it has been mentioned that strong flours take too long for conditioning and should not be used for making bread by straight dough method. For such flours sponge and dough method is more suitable where the problem of controlling the dough temperature is not so acute as the total fermentation time is divided in two separate segments. For the sake of convenience and proper identification, a sponge dough is indicated as 60/40 sponge-dough, or 70/30 sponge-dough, where the first numbers i.e. 60 or 70 indicate the percentage of flour used in sponge and the second numbers i.e. 40 or 30 indicate the percentage of flour mixed at the time of dough making.

In this method, as a first step, a part of flour, proportionate amount of water, all the formula yeast and yeast food are mixed together. Longer fermenting sponges may contain some amount of salt also. Mixing operation is carried out just sufficiently to incorporate all the ingredients evenly. This sponge is fermented for predetermined time. Sponge fermentation time depends on the amount of flour in the sponge and flour quality. The quantity of flour in sponge depends on the strength of flour. If the flour is too strong , more quantity should be used in sponge and in turn the sponge should be fermented for longer duration. It is advisable to test the sponge physically for its readiness before mixing it into dough. The following methods of sponge testing could be used; Take a small piece of sponge and try to break it with both hands. If the piece breaks with a clean fracture the sponge is ready for mixing.

If sponge is not ready, the piece will stretch to some extent and will break in unevenly stretched shreds. In such case sponge should be allowed more fermentation time. Tear the sponge apart from the center with both hands and examine the web structure. If the web structure is very fine, the sponge is ready. An adequately fermented sponge feels dry to touch without any stickiness present. When the sponge is ready, it should be broken down properly with formula water, so that its even mixing in the dough is assured. Uneven mixing of sponge in the dough should be avoided as it produces uneven results in the bread. Broken down sponge is mixed with the remaining flour, sugar, salt, fat etc. Mixing operation should be carried out to the right degree. If two different kinds of flour are at hand, the weaker flour should be used at the time of dough making.

After the dough is mixed, it is rested for 30–45 minutes during which time it relaxes from the stress of mixing operation. Pre-conditioned gluten of the sponge hastens the conditioning process of the gluten of fresh flour during this period and the dough is in perfect state for further manipulation i.e. cutting, moulding etc. This is a variation of sponge and dough method. Very often a (bread product) formula may contain milk, eggs, substantial quantity of fat and sugar as in the case of sweet bread, Danish pastry and other sweet fermented products. All these formula ingredients will have a retarding effect on yeast activity. If all the formula yeast, part of flour, yeast food and sufficient water (to make a fluid batter as in case of flying ferment) are mixed together, the yeast gets initially an environment which is conducive to vigorous activity and the end of fermentation time (of ferment) it is in a fit condition to take on the extra load of fermentation in the presence of milk, eggs, excessive fat etc.

Fermentation time of ferment depends on the formulation of the product desired to be made and the flavor desired in the product but very often it becomes a matter of individual preference eg. some bakers may take the ferment (for mixing) after it is dropped by itself , while others may take it just prior to dropping and some may allow time even after it has been dropped. A ferment containing milk should be guarded against over fermentation as it will develop more than desirable quantity of lactic acid which in turn will affect the flavour, taste and texture of the product. When a ferment is ready, it is mixed into dough, along with the remaining ingredients and allowed to ferment for the second stage of fermentation before the dough is taken up for make up. This method is used for making enriched bread, buns, Danish pastry, sweet dough, doughnuts etc. where the speed is very necessary.

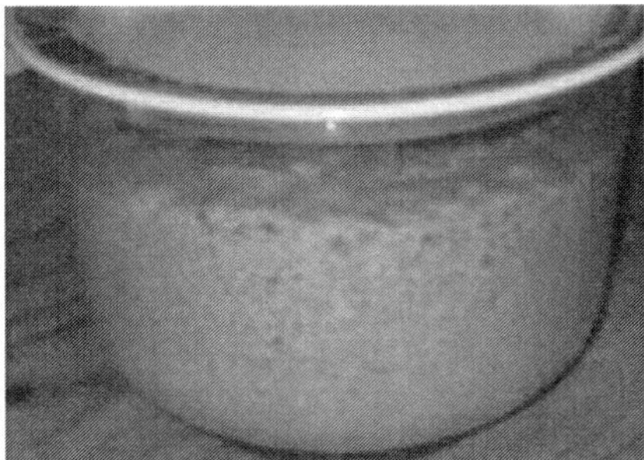

Fig 6.3 Overnight Sponge

Significance of Mixing, Fermentation and Proofing in Bread Making

Objectives of the Mixing Phase

Most of the characteristics of the final product are determined directly or indirectly during the mixing stage. If the dough is under mixed or over mixed, the handling properties of the dough will be different. Scaling is very important. If the scaling of the ingredients is wrong, the bread will show various faults, depending on which ingredient is weighed incorrectly. The importance of the dough temperature also cannot be underestimated. If the temperature isn‘t right the fermentation rate will be faster or slower, and that, in turn, will influence the volume of the bread and the color of the crust. Finally, if the mixing time is not respected, the texture and the grain of the crumb will suffer. The mixing of the dough has a number of objectives: To uniformly incorporate all ingredients, To hydrate the kur and the other dry ingredients, and To develop the gluten. To develop the gluten one has to put in energy to mix the water and kur. Slowly but surely the gluten network will start to develop. The art is to develop them to the proper consistency so the dough will have an excellent machinability as well as good gas retention properties.

Mixing Time

In a conventional spiral mixer, the mixing time for a dough of about 165 kg will be around 12 minutes, depending on such factors as the quality of the kur and the mixing method (e.g., the moment when salt is added will influence the mixing time; delayed salt addition will shorten the mixing time). During these 12 minutes one can distinguish a number of stages:

Pick up: dough is sticky, cold, and lumpy. Initial development: dough is getting warmer, smoother, and drier.

Clean up: dough is at maximum stiffness and comes together as one mass. The color will change from yellowish to a whiter, creamy color.

Final development: dough is at its correct temperature and handling quality. A gluten film can be easily obtained by stretching a piece of dough.

Letdown: dough is too warm and becomes sticky and lacks elasticity.

Breakdown: dough will begin to liquefy.

The mixing time is influenced by a great number of factors, including

- Speed of the mixer.
- Mixer design.
- Dough size in relation to mixer capacity.
- Dough temperature.
- Efficiency of cooling systems.
- Quality of the kur.
- Water absorption of the kur (infienced by the particle size).

- Amount of shortening: mixing times increase with more shortening added.
- Amount and type of reducing and oxidizing agents.
- Amount of milk solids and other dry ingredients that compete for the water: the higher their concentration is, the longer the mixing time will be because there is less water available for gluten development.

Temperature Control

During the mixing process the temperature of the dough will rise due to heat generated by the frictional forces and the heat of hydration of the kur. The frictional heat is the result of the mechanical energy one has to put into the dough in order to overcome internal and external (dough in contact with the side of the mixer bowl) friction that is caused by the dough mixing process. The amount of friction to be overcome is related to the water absorption and to the gluten development. As mixing time changes, the friction factor changes as well. The heat of hydration is the energy that gets liberated when a substance absorbs water. The amount of heat liberated varies with the degree to which water is absorbed. In the case of soluble substances, energy is needed to dissolve them, so the change in energy level is of a negative nature, as are the amounts of heat withdrawn from the system. The temperature of the dough is also influenced by other factors such as temperature of ingredients, size and type of mixing equipment (artofex mixer compared with a high-speed mixer, for instance), batch size (too small batches in too big mixers), mixing procedures (time, speed), and room temperature.

To cool down the dough and to remove the excess heat generated during the mixing process, the baker can use one of the following methods: add ice to the dough, use chilled water to make the dough, refrigerate the mixing bowl (mainly done in horizontal mixers), use a saturated salt solution which can be cooled down to below 0°C instead of granulated salt, cool down the ingredients (mainly kur, which can be easily cooled down with the injection of liquid CO_2 during the pneumatic transport).

Fermentation

Fermentation in bread making is the process by which the well-mixed ingredients for bread making are converted, under controlled temperature and humidity for an appropriate time, to a soft and expanded dough, with changes in both structural and rheological properties. The ingredients for bread making include kur, water, yeast, sugar, fats and oils, and improvers. The volume of the dough usually is expanded by several times.

The fermentation process changes the dough in two respects. First, the yeast converts the available carbohydrates (sugars) to carbon dioxide gas that enables dough volume expansion and at the same time decreases the dough pH value. Secondly, hydrolysis by the enzymes softens the gluten and changes the dough characteristics to allow more gas retention. In general, the bread making methods are classified into two broad categories: straight dough and sponge dough. The natural-sponge and boiled-sponge methods have become more popular in recent years, so these methods in bread making.

Straight-Dough Method

Among all the bread making methods, the straight dough method is the most commonly used and the simplest. In this method, all the ingredients in the recipe are put into the mixing bowl in an orderly manner and mixed into a dough, followed by a sequence of steps for fermentation, dividing and rounding, molding, final proof, and baking. Based on differences in fermentation time, the straight-dough method can further be divided into two types: straight dough and no-time dough.

Straight Dough

Straight dough is now the most widely accepted method in bakeries. Making dough through the straight-dough method implies putting all the dry ingredients such as flour, sugar, milk power, improvers, and yeast into the mixing bowl, mixing them evenly at low speed, then adding the wet ingredients such as water, ice, and egg, and continuing the process. After all the ingredients are evenly mixed, adjust the mixing speed to its medium level and continue to mix until the dough reaches the stage of expansion. Then add fats and oils, and continue to mix until the dough reaches the stage of complete mixing, which means the completion of the mixing process. The dough goes through the mixing process for only one time. The appropriate temperature range of the mixed dough is between 26 and 28 °C (78.8 and 82.4 °F), fermentation temperature is 27 °C (80.6 °F), relative humidity is 75%, and fermentation time is between 1.5 and 3 hours.

The fermentation process requires punching steps. The purpose of punching is to equalize the temperature and fermentation of the dough. The release of carbon dioxide and alcohol through punching will bring in more oxygen needed to enhance the fermentation function, accelerate gluten expansion, and increase gas retention. The exact time for punching, in principle, starts when the volume of the dough increases by one and one-half times after going through the fermentation process. Whether or not the dough has reached the time for punching can be judged based on the reaction of the dough after being pressed by the fingers. If a gentle push of the finger into the dough from the top doesn't encounter much resistance, and after the finger is removed the indentation in dough remains, the dough is ready for punch down. If the finger feels stronger resistance after pushing into the dough, and the indentation gets filled up immediately, it indicates an insufficient degree of fermentation; let fermentation continue.

Over and Under Fermentation

Fermentation stage is important in as much as the taste, volume, and keeping qualities of bread are affected when there is too much or not enough fermentation. Without proper fermentation good tasty bread cannot be made. When doughs are fermented at a higher temperature they tend to become acidic and the crumb colour has a tendency to become grayish. Bread made from over fermented doughs stales quickly. The crumb has a tendency to crumble. Over fermented doughs are inclined to become soft and sticky. They yield less bread of unappetizing appearance as quite a bit of dusting flour must be used during scaling and makeup. Over fermented doughs yield less breads and are hard to bake out. Under fermented doughs do not bake out properly and the crumb is darkish and very close. They also tend to crumble easily. Under fermented doughs have the tendency to flatten out. This can be noticed during the intermediate proof time.

Proofing

Proofing is really the stage where the production of gas in the dough is at the final stage giving volume to the bread. The ripening of the dough have been achieved during fermentation, the continued production of gas render this process complete. Average proof box temperature is normally 95—98^{0}F and humidity is 80–83%. The humidity is important to keep the crust moist and bread will not form a crust as quickly in the oven which will allow it to bloom or expand more. Proper amount of moisture is also important proper gluten conditioning. Gluten is developed and partly conditioned by mechanical mean.

Under-proofing will produce bread of small volume. At time it will brust on side and in as much the volume is less ,it will be as a rule, unbreaked. There are time when oven temperature is low that the bread should go to the oven under proofed. Over proofing to a certain degree is justified if the bread is baked in hot oven . The more open grain will allow the heat to penetrate more easily and bread will break out quickly. Excessive over proofing of bread will cause shrinkage and at time the bred may collapse in the oven. An unfermented, also known as young dough‘ will not stand as much heat than a properly fermented dough and it should be proofed less than normal. Bread made from over fermented dough also known as old dough should be full proof and if possible baked in hot oven.

Formula Construction and Computation of Yeast Raised Products

There are many different formulas for bread and yeast-raised products. Some of these formulas contain little or no enriching ingredients (eggs, fat and sugar) and would be called lean. Others have high percentage of these enriching ingredients and are referred to as rich. There are many formulas between these two extremes.

The optimum temperature ranges for the ingredients and the different phases of the dough are as follows:

1. Flour storage, 75 °F
2. Ingredient ice water, 40 °F
3. Bread doughs, 78–82 °F
4. Mixing room, 75–80 °F
5. Fermentation room or cabinet, 80 °F, 74–77 percent relative humidity
6. Proof boxes or cabinets, 95–100 °F, relative humidity 83–88%;
7. Bread slicing and wrapping temperatures 95–105 °F;
8. Inside temperature of a loaf of bread out of oven, 208–210 °F;
9. Pan temperature at time of panning, 80–110 °F;
10. Wax and cellophane wrapping paper storage, 55–85 °F, relative humidity 45–65%.

Mixing Stages

1. Early mixing stage - Rough, lumpy, wet, sticky (no trace of dough development).
2. Minutes later - Dough becomes smooth, semi-elastic, pliable, starts to dry, (start of development).
3. Further mixing - Dough attains the cleanup stage and becomes dry, elastic; does not stick to back or sides of mixing bowl (dough development).
4. After cleanup - Dough softens becomes wet, sticky and breaks short as mixing is continued (over development of breakdown).

Fig 6.4 Various types of breads

Baking

Baking is generally defined as the process in which products are baked through a series of zones, with exposure to different time periods, temperatures, and humidity conditions. As an example, in white pan bread baking.

Cold ovens

At times there is no option but to use a cool oven. In this case the bread should be underproofed so that it will not expand too much which causes shrinkage and sometimes collapsing. The routine in a bakery, especially where a direct fire brick oven is used, is to bake the small bread first which requires a hotter oven than the medium size bread and then the larger loaves, which require a cooler oven. The last to be baked is the very sweet breads, which carry a high percentage of sugar, and so forth and so on.

Hot ovens

Between a cold oven and a hot oven, one would choose a hot oven, if one has more resources. For hot oven one must ferment doughs well, use less sugar and allow the bread to proof more than normal. With a hot oven the baking is done in a shorter time and there is not the danger of having bread over proofed or doughs over fermented. The situation of cold or hot oven can happen frequently with the direct fired brick ovens wherever they are in use. Lack of proper insulation covering the dome of the oven is also undesirable. This may be sand, which is the least expensive. It should not be less than 30 inches or 75 cms thick at the top-most part of the dome. In order to do this, the wall of the oven should be build up, both front side and rear to hold this sand. The flue should be set in the chimney at the other extreme from where the firewood is burnt. Firing of an oven is very important. As a rule the oven should be fired for at least two hours. Then the flue is closed and the door shut tightly. When firing, open the door just enough to allow sufficient air to enter for proper combustion. After firing, the oven should be allowed to rest at least half an hour before starting to bake. This is done so that the heat will be evenly distributed throughout the oven and the hearth or floor is heated evenly. When firing an oven the dome and to a lesser degree the walls collect or accumulate this heat. The thicker the dome and the better the insulation on the top the more heat will be accumulated and the longer baking can be done before refiring.

Bread faults, their causes and remedy

The following gives some of the more prominent faults in white bread production:

1. Lack of volume

(a) Use of weak flour
(b) Too much salt
(c) Lack of shortening
(d) Yeast dissolved in hot water
(e) Too much or not enough dough for the mixer
(f) Under mixing
(g) Over mixing
(h) Young dough
(i) Extremely old dough
(j) Too much machine punishment
(k) Too long an intermediate proof
(l) Insufficient pan proof
(m) Excessive steam pressure in oven
(n) Oven too hot

2. Too much volume

(a) Not enough salt
(b) Use of wrong type of flour
(c) Dough slightly overaged
(d) Too much dough for pans
(e) Over proofing
(f) Cool oven

3. Crust colour too pale

(a) Too lean formula
(b) Flour lacking diastatic activity
(c) Excessive mineral yeast food
(d) Old dough
(e) Insufficient humidity in proof box
(f) Cool oven
(g) Under baking

4. Crust colour too dark

(a) Too much sugar
(b) High milk content
(c) Old dough
(d) Oven too hot
(e) Over baking

5. Blisters under the crust

(a) Young dough
(b) Excessive steam in proof box

(c) Over proofed
(d) Rough handling at oven

6. Crust too thick

(a) Insufficient shortening
(b) Low sugar content
(c) Old dough
(d) Lack of moisture in proof box
(e) Excess steam in proof box
(f) Cool oven
(g) Over baking

7. Shell tops

(a) Green or new flour
(b) Stiff dough
(c) Dough too young
(d) Lack of moisture in proof box
(e) Not enough pan proof
(f) Excessive top heat

8. Lack of break and shred

(a) Weak flour
(b) Excessive amount of mineral yeast
(c) Young dough
(d) Extremely old dough
(e) Excessive proof
(9. Crumb is grey
(a) Use of too much malt
(b) Old dough
(c) Excessive proofing
(d) Pans too large for amount of dough

10. Streaked crumb

(a) Improper incorporation of ingredients
(b) Sponge or dough crusted over during fermentation
(c) Sponge not broken up properly
(d) Excessive trough grease
(e) Scrap dough picked up during make up
(f) Excessive use of divider oil
(g) Excessive dusting flour
(h) Dough crusted during intermediate proof
(i) To much machine punishment
(j) Rough handling at oven

11. Coarse grain

(a) Weak flour
(b) Improper mixing
(c) Slack dough
(d) Young dough
(e) Old dough
(f) Improper moulding
(g) Excessive proof
(h) Rough handling at oven
(i) Cool oven

12. Poor Texture

(a) Weak flour
(b) Lack of shortening
(c) Improper mixing
(d) Slack dough
(e) Excessive trough grase
(f) Young dough
(g) Old dough
(h) Excessive use of divider oil
(i) Excessive dusting flour
(j) Improper moulding
(k) Cool oven

13. Flavour and taste are poor

(a) Improper storage of ingredients
(b) Poor quality ingredients
(c) Off-flavoured ingredients
(d) Improper amount of oil
(e) Under fermented dough
(f) Old dough
(g) Unsanitary shop
(h) Dirty pans
(i) Under-baking
(j) Over baking
(k) Bread cooled under unsanitary conditions

14. Poor keeping qualities

(a) Too lean formula
(b) Poor quality ingredients
(c) Improper storage of ingredients
(d) Old dough

(e) Stiff dough
(f) Over proofing
(g) Cool oven
(h) Bread cooled too long before wrapping

15. Holes in Bread

(a) Unbalanced formula
(b) Flour too strong
(c) Improper incorporation of ingredients
(d) Under mixing
(e) Over mixing
(f) Excessive trough grease
(g) Young dough
(h) Old dough
(l) Excessive use of divider oil
(m) Excessive dusting flour
(i) Too much machine punishment
(j) Proof box too hot
(k) Over proofing

In checking these faults an analysis of the various causes will show inferior ingredients, unbalanced formula, improper mixing, incorrect fermentation time, poor control of temperature, time and humidity throughout the production process, poor makeup procedures, poor oven conditions as well as improper handling in cooling, wrapping and shipping account for most of bread faults. A process of elimination must be instituted, the possible cause or causes determined and the proper remedy applied.

Reerences

Basic Baking. 1980. S.C. Dubey.
Edwards W.P. The Science of Bakery Products.RSC Publishing, Essex, U.K.
Gisslin, W. Professional Baking. New York: John Wiley & Sons, c1985.

7

Bakery Equipment

Various types of equipment are needed and used to facilitate the process of baking. Depending upon the use, equipment may be as light equipment and heavy equipment.

Table I: Classification of Bakery Equipment

S. No.	(i) Light Equipment	(ii) Heavy Equipment
1.	Knives - * pelliate knife * sharp knife * peeler * scraper/spatula	Oven - gas oven - coal oven/ bhatti - electric oven
2.	Scissors	Proving chamber
3.	Grater	Refrigerator
4.	Sieve	Dough mixer
5.	Strainer	Deep freezer
6.	Chopping board	Work tables
7.	Spoons - * Measuring spoons * Round spoon * Frying spoon,* Wooden spoon	Storage cabinet
8.	Rolling pin	Gas burners
9.	Whisks * Hand operated/balloon whisk * Electrical whisk	Weighing machines
10.	Measuring jug/flask	Flour sifter
11.	Enamel bowls of different sizes	Spiral dough mixer
12.	Saucepan	Planetary cake mixer
13.	Turn table	Dough divider
14.	Cake dummies	Bun divider and rounder
15.	Lemon squeezer	Bread slicer
16.	Pastry brush	Dough sheeter
17.	Weighing scales'	Rotary rack oven
18.	Sugar thennometer	Sorbet machine and Ice-cream machines
19.	Biscuit cutters	
20.	Piping bags and nozzles of different shapes	
21.	Cooling rack	
22.	Steel thalisltrays	
23.	Cake tins of various sizes and shapes	

S. No.	(i) Light Equipment	(ii) Heavy Equipment
24.	Jelly mould	
25.	Swiss roll tray	
26.	Tartlet moulds	
27.	Madeliene moulds	
28.	Flan rings	
29.	Bread moulds/tins	
30.	Baking trays	
31.	Muffin trays	
32.	Savarin moulds	

Small equipment of significance and their use

1. **Mixing bowls**: A variety of stainless steel bowls are used for whipping eggs, mixing of creams and storage purpose.
2. **Muffin pan**: The different size of baking pans with cup shaped indentation for baking muffins.
3. **Savarin moulds**: Small ring shaped doughnut shaped moulds for baking savarins.
4. **Ordinary and textured rolling pin**: Ordinary rolling pins help roll the dough and the textured rolling pins are useful to make design over the biscuit dough, and on sheets of marzipan and pastillage.
5. **Table scraper**: Use to cut pieces of dough. It is available in plastic or metals.
6. **Baking tray**: Use for baking goods. It is available in various sizes.
7. **Pastry brush**: Use to brush the items with egg wash, glaze etc.
8. **Tart pan**: Available in many shapes and sizes. They may be made in one piece or with a removal bottom to make it easier to remove the baked tart from the pan.
9. **Bread moulds**: various size of mould are used to prepare a variety of breads.
10. **Bread knife**: Bread knife is a flexible rounded tipped tool used in pastry section for spreading cream, glaze on cakes for mixing and bowl scraping.
11. **Pastry bag and nozzles**: The plastic, nylon bag is used to pipe the fillings, cream and various toppings. Nozzles are available in different designs and are used for decorating items such as cake icings and whipped creams.
12. **Sieves**: These are used to shift aerate and helps remove any large foreign raw materials from dry ingredients.
13. **Timers**: These are absolutely essential for baking.
14. **Wooden spoons**: To stir ingredients in a bowl.
15. **Juicer**: To squeeze juice from different fruits and vegetables.

16. **Hand blender**: To whisk small quantities of egg or cream
17. **Sugar thermometer**: Used to measure the temperature of the sugar or the density of the sugar syrup.
18. **Cooling wire rack**: To pull sponge cakes and yeast products and thus prevents from sweetening.

Beside these small equipments includes weighing scale, Madeleine cups, cream horn mould, chopping board, oven gloves and varies sizes of cake moulds.

Selection And Maintenance Of Heavy Equipment

1. Oven

This may be heated by coal, gas, oil or electricity. The source of heat does not matter as far as the baking products are concerned. What matters is that heat should be equally distributed and the required temperature should be available for baking. Depending upon the volume of baking, you can buy a single deck oven or a double or three deck oven. Oven can also have a proving chamber attached to it. Electric oven has a thermostat which makes temperature regulation easier and it should be possible to control top and bottom heating control top and-bottom heating control from separate knobs. In other ovens exact temperature regulation is not possible so rely more on experience for checking of temperature. Oven should be kept clean to prevent any spillage from becoming caked. Periodic checking by the electrician should be done. Ovens should be used 10–15 cm from ground to permit easy cleaning.

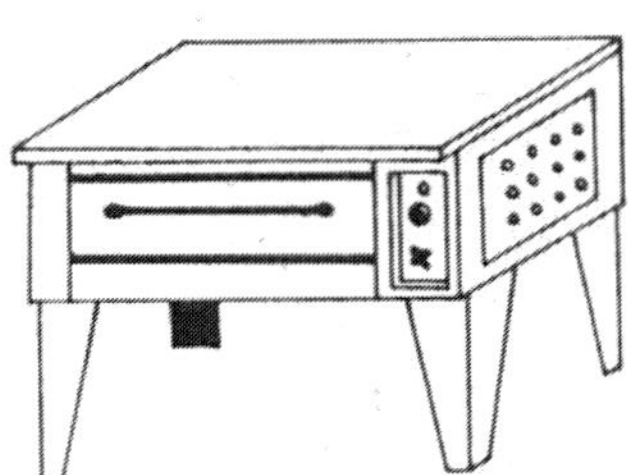

Single Deck Baking Oven

There are single, double and three decked oven available in the market. Types of oven, the product trays or moulds are placed on the oven floor. Bread baked directly on the floor of the ovens and not in pans is often called hearth breads. Deck oven for baking breads are equipped with steam injectors.

2. Proving chambers

These are cabinets with temperature and humidity control. These are used for keeping just fermented goods like bread, rolls, buns, etc., during the fermentation period so that ideal conditions can be provided for fermentation. They can be separate chambers or can be attached to the oven. These are a must for good bakers. These should be kept clean. Water should be removed and replaced after cleaning every few days.

3. Refrigerator

The capacity varies from 100 to 380 litres. The choice of a refrigerator, its overall size and the size of the frozen food storage compartment depends on the

(a) Size of the bakery
(b) Expected volume of sale
(c) Availability of other cold storage means.

It is wise to choose a larger model as per the requirement because need keeps on increasing with time. A refrigerator should be defrosted weekly. Spills should be mopped up at once and the cabinet should be washed occasionally with soda bicarbonate or detergent and water.

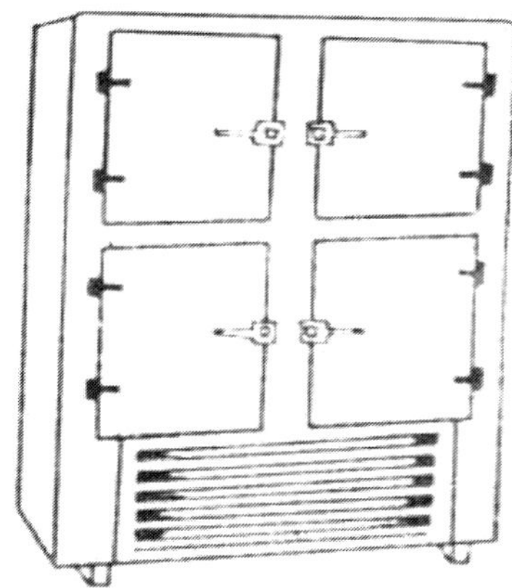

Four Door Refrigerator

4. Deep Freezers

The capacity varies from 140–380 Ltrs. It can be of three types:

(a) The chest type with top opening lid
(b) The upright type with front opening
(c) A combination type with both openings

Freezers should be defrosted at least once a month. Wash the insides with soda bicarbonate and water and dry before switching it on. Most freezers and refrigerators work for years unimpaired but servicing may be required.

Deep Freezer/Ice Cream Cabinet Bottle Coller

5. Dough Mixer

It is generally made to order but available capacities vary from 25–35 kg. Most commercial modules are heavy and should be fitted on sturdy rollers for easy movement. The stainless steel bowl and beater should be washed and cleaned after every use and the machine should be serviced regularly.

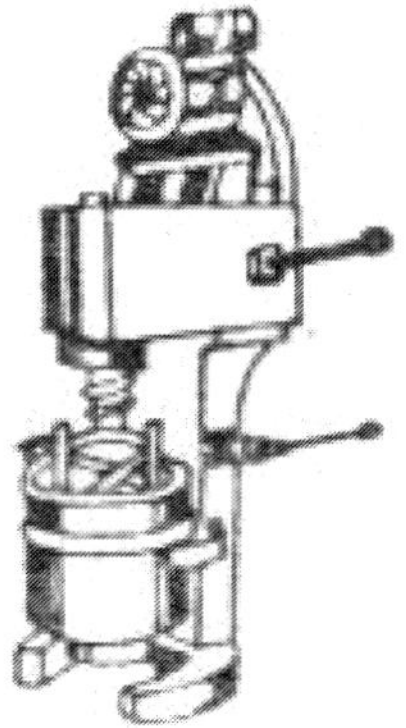

Dough Mixer

6. Work Tables

Steel preparation tables last for years. The table top could also be made from marble which is smooth, easy to keep clean and remains cool. There should be no unwanted cracks or joints and the design should be simple. Tables with open sides and without drawers are the best as dirt does not accumulate.

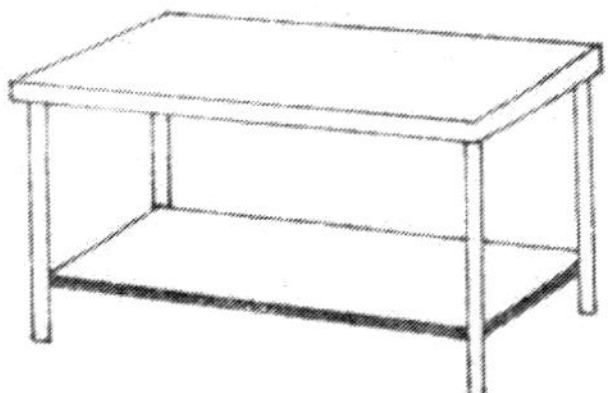

Work Table with Bottom Shelf

Work table should be maintained scrupulously clean as they can cause cross contamination. They should be scrubbed with plastic brush and detergent and washed and dried.

7. Storage Cabinets

These should be good sized food cupboards to store provisions and smaller equipments, some of them should have air-vents for proper ventilation. These can be built-in types or made of metal-free standing style. But all cabinets should be kept clean and free from pests.

8. Grass Burners

One low pressure gas burner with a simple and easy to clean design is a must in a bakery. Preference should be given to stainless steel instead of enameled metal. Gas ranges should be periodically serviced to check on burner's efficiency and they should be kept clean and free from obstruction.

9. Weighing machines

Raw material measurement used in proper weighing scale is very important for the quality product for the accurate quantity.

Weighing Machine

10. Flour sifter

Flour sifter is an essential part of food safety system (HACCP). It will aerate the flour and other ingredients for getting better volume of finish products.

11. Spiral dough mixer

Spiral dough mixer is a specially design for making large quantity of yeast dough. There are two models of mixers available in the market. Most models have a single vertical mixing arm or hook. Another model machine is having two agitator arms which are mounted vertically on circular poles.

Spiral dough mixer

12. Planetary cake mixer

Two types of mixer are available in the market:

1. Bench model
2. Floor model

These mixers usually have three operating speeds and their mixing attachments, namely

1. Wire whip
2. Flat beater
3. Dough arm

Bench model planetary cake mixer

Floor model planetary cake mixer

13. Dough divider

The dough divider machine divides the bulk dough into desire size. The dough density should be even otherwise the weight might change. Single pocket divider will be easier to use.

14. Bun divider and rounder

Bun divider and rounder divides the dough into many pieces at once and it then automatically rounds all of them, greatly speeding the makeup of the dough products.

Bun divider and rounder

15. Bread slicer

Gravity feed slicer are best suited to the small, wholesale and large retail bakeries where a great number of sliced breads are produced. All types of white and sweet bread can be sliced without wastage or damage. The cut slices come out at the output end.

Bread slicer

16. Dough sheeter

A dough sheeter rolls out portion of dough into sheets of uniform thickness. The machine consists of a canvas conveyer belt that feeds the dough through a pair of rollers.

Dough sheeter

17. Rotary rack oven

A rack oven is a large oven into which entire racks full of sheet pans can be wheeled for baking. They are also equipped with steam injectors. Rotary rack ovens are excellent for large scale production. It can be fired with gas or electricity.

Rotary Rack oven

18. Sorbet machine and Ice-cream machines

Sorbet machine is used to churn puree along with sweeteners and other flavors. Ice cream machine is used to make frozen dessert topping.

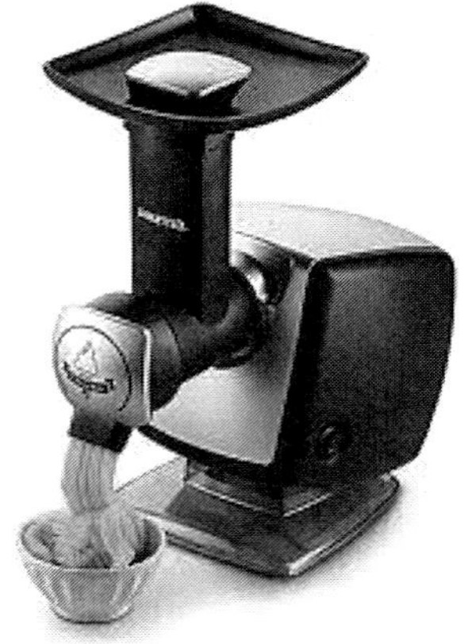

Sorbet Machine **Ice cream machine**

References

Cereal Processing Technology, ed. G. Owens, Woodhead Publishing, Cambridge, 2001.

Manual for Small Scale Bakery Units, FSSAI.

Sultan, W.J. Practical baking. 5th edition. New York : Van Nostrand Reinhold, 1990

8
Quality Assessment and Standard Specifications of Bakery Products

The raw material of foremost importance in bakery product is the wheat flour. Bakery units prefer the flour obtained by milling in roller flourmill with 70–72 percent extraction. Flour quality may be defined as the ability of the flour to produce an attractive end product at competitive cost, under conditions imposed by the end product manufacturing unit. The concept of quality differs from producer and consumer point of view. However, in general, the term quality may refer to fitness of a raw material or a product for a particular process or consumer. For a consumer, the following parameters are important criteria of a product quality.

1. Uniformity and consistency of quality
2. Health safety of the product and
3. Price

The tests most commonly used to predict the quality of wheat flour and bakery products are described as follows

1. Moisture Content

Principle

The moisture content is the loss in weight of a sample when heated under specified conditions.

Scope and objective

It is applicable to flour, farina, semolina, bread and wheat grain. Flour moisture is influenced by weather and environmental or storage conditions such as humidity and storage temperature. Such conditions affect the keeping quality of flour. Higher moisture may lead to spoilage and lump formation during storage. Lower moisture content, on the other hand, cause loss to the baker in terms of low dry matter. Several methods are available to determine moisture content e.g. air oven method, direct distillation, chemical and electrical methods. In air over method 5 gm sample is kept in a dish for one hour at 130°C. Electrical method could also be used satisfactorily provided they are accurately calibrated.

Apparatus

1. Wiley Laboratory Mill, intermediate model, equipped with 18 or 20-mesh screen or any other mill that will grind to same degree of fineness without under exposure to atmosphere and without appreciable heating.
2. Oven (either gravity-convection or mechanical convection). Capable of being maintained at 130°C (+1°) and provided with good ventilation. Thermometer shall be so situated in oven that tip of bulb is level with top of moisture dishes but not directly over any dish.
3. Moisture dishes having diameter of 55 mm. and height of 15 mm. They should be of heavy- gauge aluminum with slightly tapered sides and provided with tightly fitting slop in covers. Before using, dry for 1 hr. at 130 °C, cool in desiccators, and obtain tare weight.
4. Airtight desiccators containing activated alumina.
5. Balance, accurate to at least 1 mg.

Method

1. Grind a 30–40 g sample in mill, leaving minimum possible amount in mill. Mix rapidly with spoon or spatula and transfer immediately a 5 g portion to tared moisture dishes. Cover and weigh dishes at once. Subtract tare weight. And record weight of sample. Dismantle and clean mill between samples.
2. Uncover dishes and place them with covers beneath on shelf of oven. Insert shelf in oven at level of thermometer bulb. Heat for exactly 60 minutes after oven recovers its temperature of 130 °C.
3. Remove shelf and dishes from oven, cover rapidly (using rubber finger insulators), and transfer to desiccators as quickly as possible. Weigh dishes after they reach room temp. (45–60 minutes, usually). Determine loss in weight as moisture. Replicate determination must check within 0.2% moisture.

Calculation

$$\text{Moisture (\%)} = \frac{(A-B)}{(A-C)} \times 100$$

Where, A = wt. of flour + Aluminium dish before drying

B = wt. of flour + Aluminium dish after drying
C = wt. of aluminium dish

Standard values in flour

ISI ... 13.0%; PFA ... 14.0%

2. Protein Content

Principle

Protein in wheat flour and bakery products is generally measured using the Kjeldahl method. This method estimates the total nitrogen in a sample and assumes a constant relationship between total nitrogen and the protein in wheat. The results are expressed by multiplying the nitrogen content by 5.7 factor and hence this method is reported to measure 'crude protein'. More recently, methods have been developed to determine protein quantity by near-infrared reflectance (NIR) technique. Wheat flour protein quality is difficult to estimate, as there is no standard method available so far. However, some methods such as sodium dodecyl sulfate sedimentation test, Pelshenke test and extensibility test on wet gluten using Texture analyser are employed to assess the quality of wheat flour for specific product.

Method

1. Place 1g sample in digestion flask. Add 0.7g HgO or 0.65g metallic Hg, 15g powdered K_2SO_4 or anhydrous Na_2SO_4, and 25 ml H_2SO_4.
2. Place flask in inclined position and heat gently until frothing ceases. If necessary, add small amount of paraffin to reduce frothing. Boil until solution becomes clear.
3. Cool to 25 °C and add 200 ml distilled water. Then add 25 ml of sulfide or thiosulfate solution and mix to precipitate Hg. Also add few Zn granules to avoid bumping, tilt flask and add NaOH without agitation.
4. Immediately connect flask to distilling bulb on condenser and with tip of condenser immersed in standard acid and 5–7 drops indicator in receiver. Rotate flask to mix contents, and then heat until all NH_3 had distilled.
5. Remove receiver, wash tip of condenser and titrate excess standard acid in distillate with standard NaOH solution. Correct for blank determination on reagent.

Calculation

% Nitrogen (N) = [(ml standard acid normality acid)
– (ml standard NaOH normality NaOH)] × 1.4007/g sample
Multiply % N by 5.7 to get % protein.

3. Ash Content

Principle

Total ash is the inorganic residual remaining on incineration in a muffle furnace. This reflects the quantity of mineral matter present in the flour. Acid insoluble ash reflects added mineral matter in milled products such as dirt, sand, etc.

Objective

Ash, an index of the mineral content of the flour, gives an indication of the grade or the extraction rate of the flour. This is because the mineral content of the endosperm is very low, as compared to the outer bran layers. Thus, low-grade flours, rich in powdered bran give higher ash contents as compared to more refined or patent flours.

General method

Weigh 10 g of the sample into a weighed silica dish. Incinerate it over a burner or in the muffle. Keep the dish in a muffle furnace maintained at 550–600 °C until light grey ash results or to a constant weight, cool in a desiccators and weigh.

Rapid method

Reagent

Alcoholic Magnesium Acetate Solution

Dissolve 15 g Magnesium Acetate Tetra Hydrate (Mg $(C_2H_3O_2)$ 4 H_2O) in alcohol and make up to 1 litre.

Determination

Weight 10 g of flour into a weighed silica dish. Add 10 ml. of the reagent. Let the mixture stand for about 2 minutes. Evaporate the excess alcohol in a water bath and keep in muffle furnace maintained at 750 °C–850 °C for 30–45 minutes. Remove the dish, cool in a desiccators and weigh. Determine the blank on 10 ml of the solution. Deduct blank from ash.

Acid insoluble Ash

Boil ash obtained in method 1 with 25 ml HCl (1: 2.5) for 5 minutes on a water bath, covering the dish with watch glass. Filter through ash less filter paper (No. 40). Wash the residue with water until free of acid. Ignite at 600 °C for 20 min, cool and weigh.

Calculation

$$\text{Ash} = \frac{W_3 - W_1}{W_2 - W_1} \times 100$$

Where, W_1 = Wt. of silica dish
W_2 = Wt. of silica dish + sample
W_3 = Wt. of silica dish + ash

$$\text{Acid insoluble ash (\%)} = \frac{W_4 - W_1}{W_2 - W_1} \times 100$$

Where, W_4 = Wt. of silica dish + acid insoluble ash.

PFA limits: On dry basis	Atta	Maida
Ash (%)	2.00	1.00
Acid insoluble ash (%)	0.15	0.10

4. Minerals Estimation

Principle

The mineral content of flour as such is not related to quality of a final product, but is does affect the appearance of flour and the product. The minerals are concentrated on the outer part of wheat grain, which is removed during milling. However, some contamination does occur in flour. Flour that contains higher proportion of minerals will have more ash content and it will be darker in colour and it may also contain more fine bran particles. Bran has been shown to have detrimental effect on the quality of bakery products.

5. Diastatic Activity and Maltose Value

Principle

The diastatic activity is the test, which reveals the extent to which the diastatic enzyme alpha-and beta-amylases produce sugars while acting on starch present in the flour. Normally, wheats have sufficient beta-amylase activity but lack in alpha-amylase activity. However, amylase activity increased thousand folds during wet harvest or germination. The diastatic activity is expressed as mg maltose produced/10 g of flour in one hour at 30 °C. The optimum level is between 2.5–3.5 (150–350 mg/10.0 g flour). It has been reported that the flours with maltose figure of less than 1.5% or 150 mg maltose/10g may tend to be deficient in gassing power. On the other hand, when the maltose figure is over 2.5% (250 mg per 10 g. flour), there is a danger of excess gas production so certain amount of diastatic activity in flour is most essential for bread making. For cookie and biscuit making, high diastatic activity is not desirable and the flour unfit for bread-making purposes due to low diastatic activity can easily be used for cookie/biscuit making.

Reagents

1. 1:5 Sulphuric acid (200 ml concentrated sulphuric acid made unto 1 litre).
2. 15% solution of sodium tungstate
3. Fehling's Solution A: Weigh accurately 69.28 g of copper sulphate and make up the volume to 1 litre with distilled water and filter.
4. Fehling's Solution B: Weigh accurately 346g of sodium potassium tartrate and 106g of sodium hydroxide pellets and make up the volume to 1 litre with distilled water. The solution is kept overnight and filtered through glass wood.
5. Methylene blue: 1% solution in distilled water.

Method

Place 15g of flour in a 250 ml dry bottle and add 15 ml of water at 27 °C. Keep the bottles with the contents at 27 °C for one hour, the contents of the bottle being mixed by shaking once every 15 minutes during this time. At the end of the digestion period, add 1.5 ml of 1: 5 H_2SO_4 (Reagent No. 1) and 3.5 ml of sodium tungstate (Reagent No. 2) to stop the reaction. Filter immediately through No 1 filter paper and the clear filtrate is used for the determination of sugar content.

Take Fehling's solution A (5 ml) and Fehling's solution B (6 ml) in a 250 ml conical flask. Place the flour extract in a 50ml burette. Heat the mixed Fehling's solution on a burner and run at least 15–20 ml of flour extract into the flask. Add 5 drops of methylene blue, heat to boiling and continue boiling for one minute, then add additional extract slowly at a time while still boiling until the blue colour disappears. An extra drop of indicator is helpful at the end. Repeat the titration. Calculate the maltose value from the Table 1. The maltose figure of flour sample should preferably be between 1.5 and 2.3. 11

Table 1. Maltose figures corresponding to various titration levels

Flour extract (ml)	Maltose	Flour extract (ml)	Maltose
15	3.61	33	1.61
16	3.38	34	1.56
17	3.18	35	1.52
18	3.00	36	1.47
19	2.84	37	1.45
20	2.60	38	1.40
21	2.56	39	1.36
22	2.44	40	1.32

Flour extract (ml)	Maltose	Flour extract (ml)	Maltose
23	2.33	41	1.29
24	2.33	42	1.26
25	2.14	43	1.23
26	2.06	44	1.20
27	1.99	45	1.17
28	1.91	46	1.15
29	1.84	47	1.12
30	1.77	48	1.10
31	1.72	49	1.08
32	1.66	50	1.05

6. Estimation of Gluten Quantity

Principle

To separate gluten from other constituents, the wheat flour is mixed with water. The native proteins of flour interact to form a chewing gum type of wet mass, which is called wet gluten. The wet gluten can be washed out using potable water using automatic gluten washer. The wet gluten is dried to form a free flowing light coloured powder. Depending upon the variety, it has been noticed that wide variation in the quality of extracted gluten occurs.

Scope and objectives

The procedure is applicable to whole-wheat meal and refined flour. The dough developed by mixing wheat flour with water possesses the visco elastic characteristics vital for dough handling and final product quality. The visco elastic nature of dough is attributed to gluten proteins namely gliadins and glutenins. The gliadins impart extensibility to dough, whereas glutenin is held responsible for strength and elastic character of gluten and dough.

Method

Quantity of wet gluten is estimated using automatic gluten washer. The equipment has a mixing chamber. The bottom of the chamber holds an 80μm sieve. The sieve is moistened before use to achieve a capillary water bridge that prevents flour loss. A 10g flour sample is developed into dough using 5.2 ml of 2% sodium chloride solution and the dough is introduced into the plastic chamber of the gluten washer. Washing is started and after 10 min of washing cycle with 2% sodium chloride solution the washing cycle completes. The wet gluten so obtained is weighed and it is flattened between twin hot plates of the drier, where it is heated for 4 min. The dried, thin sheet of gluten is then weighed and recorded as dry gluten. The wet gluten can also be dried in oven at 100°C for 24 hours to get value of dry gluten.

Calculation

$$\text{Wet gluten (\%)} = \frac{A}{C} \times 100$$

$$\text{Dry gluten (\%)} = \frac{B}{C} \times 100$$

Where, A = wt. of wet gluten; B = wt. of dry gluten and $\underline{C}$ = wt. of flour.

7. Sds-sedimentation Volume Test

Principle

The wheat flour is treated with lactic acid and sodium dodecyl sulphate (SDS) solution. SDS neutralizes the charge of proteins and lactic acid makes the protein strands swell in the solution. Depending upon the quantity and composition of gluten proteins the flour gives some value of sediment. The volume of sediment formed when flour is suspended in water containing lactic acid and SDS is referred to as SDS-sedimentation volume.

Reagents

1. Lactic acid solution: 3 ml of 88% lactic acid is diluted (1: 8 v/v) to 27 ml with distilled water.
2. SDS solution (2%): Dissolve 20 g SDS (Sodium dodecyl sulphate "Specially pure") in distilled water to make 1 litre.
3. Lactic acid-SDS stock solution: Add 20 ml of Reagent (1) to (2).

Procedure

The sodium dodecyl sulphate (SDS) sedimentation volume of flour samples is estimated according to the method of Axford (1978). Flour (5g, 14% moisture basis) is added to water (50 ml) in a cylinder, a stop clock is started and the material dispersed by rapid shaking for 15s. The contents are re-shaken for 15s at 2 min and 4 min. immediately following the last shake, SDS-lactic acid reagent (50ml) is added, and mixed by inverting the cylinder four times before re-starting the clock from zero time. The SDS-lactic acid reagent is prepared by dissolving SDS (20 g) in distilled water (1L) and then adding a stock diluted lactic acid solution (20 ml; 1 part lactic acid plus 8 parts distilled water by volume). Inversion (four times) is repeated at 2, 4, 5 and 6 min before finally starting the clock once again from zero time. The contents of the cylinder are allowed to settle for 40 min before reading the sedimentation volume.

8. Falling Number Test

Principle

The Falling Number test (AACC Approved Method 56-81B) provides an index of α-amylase in a flour or ground-wheat sample. The procedure relies on the reduction in viscosity of starch paste caused by the action of α-amylase. The method is based on the unique ability of alpha-amylase to liquefy a starch suspension. Gelatinization strength is measured by falling number as "time in seconds" required stirring and allowing the stirrer to fall a measured distance through hot aqueous flour gel undergoing liquefaction.

Method

1. The distilled water in bath is brought to boil.
2. Weigh 7 gm of flour, transfer it to viscometric tube, and add 25 ml of distilled water, rubber the tube and shake vigorously for obtaining a uniform suspension.
3. Remove stopper and push down flour adhering to sides with the viscometer stirrer.
4. Place the tube with stirrer in the boiling water bath. Start the timer.
5. After 5 seconds, automatic stirring starts at the rate of 2 stirs/seconds for 60 seconds. After a total of 60 seconds stirring automatically stops releasing the stirrer at its uppermost position and allows falling by its weight at a fixed distance and time is recorded in seconds. The starch gelatinizes, and the α-amylase liquefies the resultant paste. The time it takes (in seconds) for the viscometer stirring rod to fall through the starch paste is the Falling Number. Flour made from sprout-damaged wheat can have a Falling Number of 100 sec or less. Bread wheat with average α-amylase activity has a falling Number of approximately 250 sec. The upper limit for the Falling Number test is approximately 400 sec, which occurs for flour devoid of α-amylase.

Interpretation of results

Falling number (in seconds)
Below 150 - sprouted wheat, high alpha-amylase activity.
200–250 - normal alpha-amylase activity.
300 and above - amylase activity too low.

10. Dough Raising Capacity

Principle

Yeast is a biological material, and thus its activity is affected by many factors such as storage temperature, relative humidity and moisture content, etc.

Such conditions affect the number of viable cells per unit mass and hence the dough raising capacity. In order to produce good quality fermented product it is important to add the optimum quantity of yeast. Therefore, it becomes necessary to check the dough raising capacity of each batch and also periodically for satisfactory gas production during fermentation of the product.

Method

Yeast (2.5 g) is dissolved in water (45ml) having 40 °C temperature. Wheat flour (35 g) is taken in a beaker, 1g sugar is added to it and then mixed with the yeast suspension. This mass is made into smooth batter and transferred to a 250 ml graduated cylinder and base volume of the batter is noted down. The rise in the level of dough is noted at 15 minutes interval for one hour. A graph between time and the rise in dough volume is plotted to estimate the dough raising capacity of yeast.

Calculation

$$\text{Dough raising capacity} = \frac{(B - A)}{A} \times 100$$

Where, A = volume of the dough before fermentation.

B = volume of dough after one hour fermentation.

11. Alkaline Water Retention Capacity

Principle

The alkaline water retention capacity (AWRC) is the amount of alkaline water retained by flour at 14% moisture under controlled centrifugation condition. The test is actually the weight of the 0.1 N sodium bicarbonate held by a flour sample following centrifugation. The gain in weight is expressed as per cent alkaline water retention capacity of flour.

Method

Flour (1 g) is slurried with 0.1 N sodium bicarbonate (5 ml). It is then shaken, allowed to hydrate for 20 min, and centrifuged under specified and constant time and centrifugal force conditions. The supernatant is decanted, the weight of the wet flour is determined, and AWRC is calculated. This parameter is important when the water relationships in a product are critical to product quality. One specific application of this test (AACC Approved Method 56-10) is as a flour specification to predict cookie spread. As AWRC increases, cookie spread decreases.

Significance of Above Functional Tests in Relation to Bread, Biscuits and Cakes

Several tests are carried out to predict the end use quality of flour. These tests are classified as physiochemical, functional and rheological tests. The functional tests such as gluten quantity, SDS-sedimentation volume, falling number, dough raising capacity of yeast and alkaline water retention test are performed to judge the quality of raw material particularly flour to get best potential of a flour when processed at industrial scale. These tests are useful in making compatible application of flour for a specific product. This avoids processing losses and helps in improving the overall quality of product. Various classes and varieties of wheat with diverse technological significance are grown in India. Thus, functional tests are carried out for using the right variety for a specific product. The technological potential of wheat is attributed mainly to its gluten proteins. Thus, quantity and quality of gluten proteins are assessed for the industrial application of wheat variety. The supplementation of wheat gluten in weak flour has potential to transform poor quality wheat into good quality, which can be processed into any value added bakery product. Wet gluten for good bread flour falls in the range of 30–36%. Flour having wet gluten of 22–25% is suitable for biscuit and cookies production. Dry gluten for good bread flour falls in the range of 10–12%. For soft wheat products the range of dry gluten should be 8-9%.

SDS-sedimentation volume test gives indirect measure of quantity and composition of gluten proteins. Higher sedimentation volume reflects appreciable quantity of high molecular weight glutenin proteins in the flour. Such flour is recommended for the bread making. Flour with lower sedimentation volume, on the other hand, is preferred for biscuit/cookie or cake production. Sedimentation value for different application of flour is as follows:

1. Soft wheat flour having sedimentation volume less than 20 ml should be used for sweet biscuit/cookie/cakes.
2. Medium strong wheat flour having sedimentation volume from 20–40 ml is preferred for fermented biscuits.
3. Hard wheat flour having sedimentation volume more than 40 ml is found useful for sbread formulations.

Falling number indicates activity of amylolytic enzymes of a flour. Flour should have desirable enzymatic activity for superior quality of a product. Wheat flour should have different enzymatic activity for processing flour in to different product. Because α-amylase hydrolyses starch link-ages, more free sugars are liberated and lower starch paste viscosity results when enzyme activity is high. The implications of this for a baked product can be very significant because of the functional roles that starch plays in most products. High α-amylase activity can lead to excessive browning because the reducing sugars liberated are available for Maillard browning reacting. Reduced viscosity caused by α-amylase can have devastating effects on batter products, reducing volume and producing an undesirable crumb structure. In bread products, hydrolysis of starch can lead to sticky crumbs and reduced volumes as well. Falling number of Indian wheat flour are very high with a mean value of 571 which indicate no alpha-amylase activity. The bread produced from such flour will have diminished bread volume and dry crumb. On the other hand, rain soaked wheat contains a very high alpha-amylase with falling number value even less than 100 which produce bread with sticky crumb, dark crumb colour and low volume. Ideal falling number value for bread making should be 175–275.

The dough raising capacity of yeast is useful observation for fermented products such as bread, buns and fermented biscuits. The yeast being a biological material is affected by storage conditions. Dough raising capacity indicates viability of yeast and hence helps in assessing the optimum quantity of yeast to be used for a product. In dough raising capacity test, the yeast (compressed) shall be deemed to have satisfied the test, if the rise in dough level is at least 110% of the original dough level. For dry yeast and the rise in dough level should be at least 80% of the original dough level. Alkaline water retention capacity (AWRC) test is useful for cookies. The results of alkaline water retention capacity directly correlate with cookie diameter, since the test is done at pH 8.0–8.1, the conditions that actually exist in cookie doughs. It is much more informative, when combined with baking. The higher AWRC value indicates smaller cookie diameter and vice-versa.

Texture assessment of Bakery products

There is a very narrow range of textural characteristics that are associated with bakery products — consumers expect certain things when it comes the way these products feel and behave when eaten. Any variance from the expected texture is considered a dip in quality. The difference in the way certain textural properties in bakery products are viewed and the way they are associated with quality are described below:

- Moist cakes and hard biscuits are regarded as quality products.
- Bread is expected to be soft to a certain degree.
- Moist bread is considered stale.

Raw materials and components used in automated bakeries play an important role in manufacturing high quality bakery products. When these are are sold to distribution networks, storage conditions need to be considered as they can affect shelf-life. The quality of the dough and the raw ingredients used in the bakery can be tested for adherence to industry standards with the help of the FTC's range of fixtures for dough.

Test Methods for the Bakery Sector

The texture of bakery products can be tested with a number of texture testing techniques. These techniques are practical ways of checking the texture based on the way they are handled by the consumer, with tests including breaking, bending, stretching, cutting and squashing. The acceptability of bakery products can be tested by emulating the way they are handled, for example, the freshness of bread can be tested by squashing it. It is also important to consider that the coatings or fillings used on baked products may have considerable effect on the texture of the final product when it reaches the consumer.

Texture analyzers, or texturometers, are instruments used to apply scientific methods to the measurement and analysis of product texture. Such equipment is used to evaluate and control the factors affecting quality, processing, handling and shelf-life, enabling the producer to understand consumer habits and acceptance criteria. Food texture, through touch, is very important factor for the end customer, in addition to taste and smell. Cosmetics and pharmaceutical products are also sensory-dependent and benefit greatly from a texture analysis capability.

The principle of a texture measurement system is to physically deform a test sample in a controlled manner and measure its response. The characteristics of the force response are as a result of the sample's mechanical properties, which correlate to specific sensory texture attributes. A texture analyzer applies this principle by performing the procedure automatically and indicating the results visually on a digital numerical display, or screen.

Forces created during this movement are manipulated to recreate consumer interactions, for example the conditions that foods are exposed to when eaten or processed. This enables the direct measurement and thus the ability to predict how a product will perform or feel. There are fundamental test methods designed to deform samples in ways which simulate these complex human interactions and replicate such conditions.

Benefits of the texture analyzer

Instrumental methods of assessing texture can be carried out under strictly defined and controlled conditions. In food production, changes in ingredient levels can cause several simultaneous changes in product characteristics. A texture analyzer has the capacity to replace human sensory evaluation by providing a numerical value which serves as a quality assurance standard by which products can be evaluated.

Reliable, repeatable, accurate testing – consistent company-wide

- Quality assurance – confidence in product performance
- Processing equipment settings optimization – mixing, transport, storage and packaging
- Respond to new regulations or legislation
- Create new and alternative products with desired texture
- Cost savings – maximise efficiency

Different test applications call for different levels of technology, load capacity, robustness, accuracy and budget. These requirements directly affect the choice of texture analysis apparatus.

The principle of texture instruments

Texture analyzer measures the texture of a sample. This instrument measures time, force and distance in the same time to describe physical characteristics and draws the texture curve from this three data and calculates the result from it.

Measure Force

Force measurement is through the loadcell. First, Force is converted to deformation by sensor. The sensor converts deformation to analog electrical signals. The analog signal will transfer to digital data using IC and internal program. The quality of the sensor also plays a role in determining, which is important for the subsequent stability and accuracy of the data.

Measure Distance (Displacement)

Displacement measurement also plays a very important role. High-precision motors include stepper motors and servo motors. Both motors accurately control the distance and deliver the distance parameters instantly using the instrument's internal procedures. The motors need high precision, small vibration (to reduce the impact of loadcell), low noise (high quality instrument) features.

High-precision ball screw

When the motor turns, the arm moves up and down using the high-precision ball screw. The probe is fixed on the arm. Sensor will measure the force from sample.

Time, force and distance of the three data is used to draw graph

Force, time and distance can be continuously collected to original data (Raw Data). The data point of original data is determined according to the experimental requirements. In principle, the more data points, the more precise curve can be obtained, and the more immediate the stress change of the sample can be reflected. But the greater the number, the more noise it collects at the same time, and it also burdens associated hardware.

Raw Data - Data format

Time	Distance	Force
1	0	0
2	1	10
3	2	30
4	3	60
5	4	100
6	5	150
7	4	95
8	3	55
9	2	32
10	1	8
11	0	2

Three common texture curves that can be generate using texture analyzer

1. **Force vs. Time curve:**

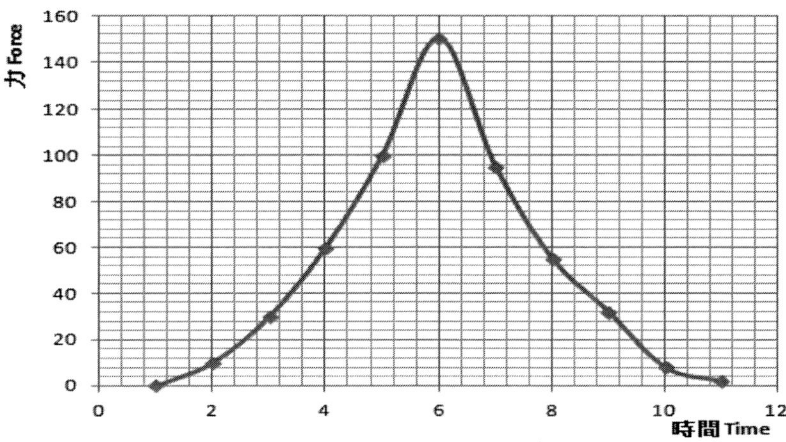

Force vs. time curve, this plot is the most important model, the user can clearly see the sample shape and force change over time.

2. Force vs. Distance curve:

Force vs. Distance curve, is also a common way of drawing, but compared to the first way of drawing, less used. This method can directly show the force change versus distance change, it also can understand the elasticity of the sample.

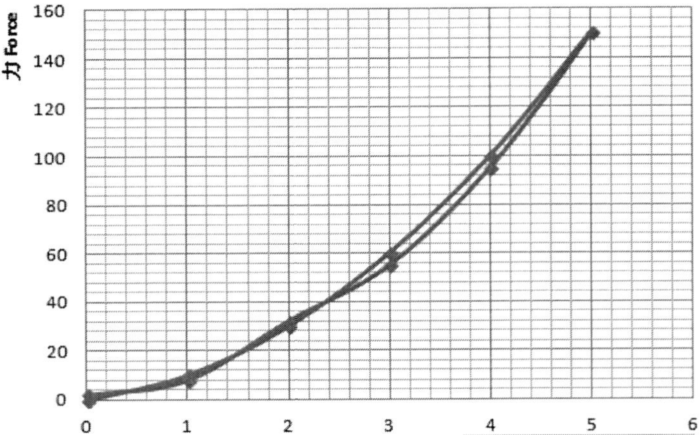

3. Distance vs. Time curve:

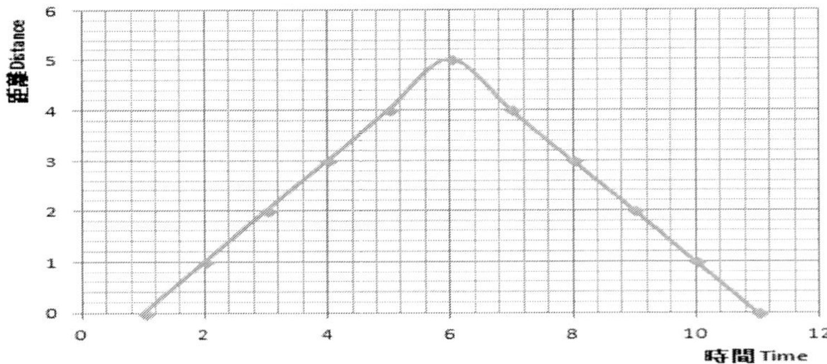

This plotting method is relatively rare, mainly used to confirm the speed and distance settings are correct, only a small number of opportunities will be used.

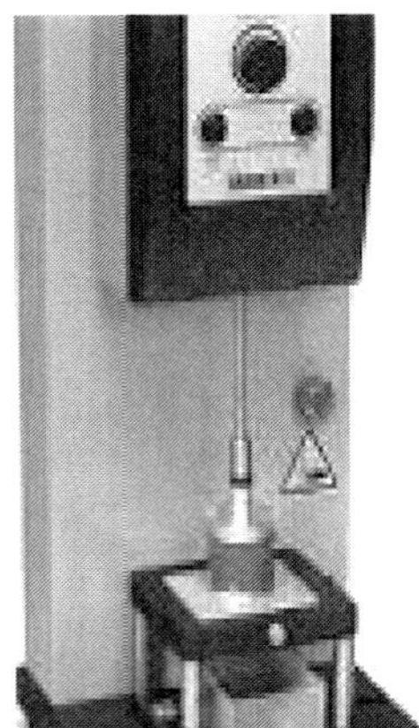

Texture analyzer (Brookefield CT3)

Some of the common testing techniques are:

- Deformation assessment of bakery products by subjecting them to shearing, compression and penetration
- Hardness assessment through breaking tests and bulk analysis

These techniques are made possible by different probes that come as accessories along with texture analyzer

Bulk analysis

Crouton hardness is measured through bulk analysis. Bulk analysis is suitable for when testing a single sample is not feasible, or when it would not represent the way it is handled by the customer. The shearing, compression and extrusion actions during consumption are replicated by the Kramer Shear Cell (Figure 1). The effect of all these actions is measured together, enabling increased reproducibility of results for a highly variable product. This method can be used for measuring the crunchiness and freshness of biscuits, and the integrity of the fruit used for filling pies.

Fig. 1 Measuring Crouton hardness via bulk analysis using Kramer Shear Cell

Compression

The most appropriate indicator of the freshness of the product is compression, for instance, the way in which a loaf of bread recovers after compression is indicative of its quality. If it retains a good amount of springiness it can be proven to be fresh, whilst crumbliness is a sign of excess loss of moisture. The deformability can be tested through FMBRA (The Flour Milling and Baking Research Association) compression testing. Many textural properties like cohesiveness, springiness, hardness and fracturability can be assessed by squashing solid and self-supporting samples. This method can also be used for conducting the texture profile analysis of the breadcrumb texture.

Fig. 2 FMBRA compression testing

Extrusion (back extrusion)

The back extrusion test (Figure 3) can be used for measuring the flow and consistency of custard. This method is also suitable for testing softer food materials like liquids and pastes that are usually tested inside their packaging. Based on the controlled manner in which semi-solid and viscous liquids are displaced, properties like thinning, consistency, viscosity, spreadability, flow and adhesiveness can be assessed. The back extrusion method can be used for comparing the composition of different custard formulations, the spreading and yielding properties of fillings and sauces, and the consistency of batter.

Fig 3 Back Extrusion test

Penetration and Puncture

To test the firmness of bread, large size cylinders are pressed into the center (Figure 4), while small balls, cylinders, cones and needles are penetrated into the surface of the sample to replicate the biting action. The changes in the crispness and hardness of bakery products can be tested through penetration which helps in assessing new formulations, or how different conditions affect shelf-life.

Fig 4. Firmness testing of bread

Multiple Point Penetration

This testing method helps evaluate the firmness, maturity and gel strength at multiple points on the same sample since the form and texture may be different in different sections.

Fig 5. Multiple point penetration testing

Shear

Bakery products like croissants need to be subjected to shearing in order to assess their toughness. Cross sections of croissants are sliced with blades and wires to replicate the action of the front incisor teeth for assessing their tenderness, bite strength and toughness.

Fig 6. Shear testing of fruit cakes

The sample is subjected to tearing, shearing and compression based on the geometry of the blade, while the texture variations in the sample are evaluated by slicing through it. This method can also be used for evaluating the adhesiveness and consistency of particulates present in fruit cakes. The texture changes due to various processing methods of short dough can be assessed by cutting through it. In order to optimize mixing properties, the cross-sectional hardness of bakery fats and butter can be measured as per ISO 16305.

Snap, Bend and Break

The fracture characteristics of biscuits can be assessed by using a three point bend test Figure 7). The fracturability of brittle solids in the form of bars can be tested by this method. The sample is supported at both ends and a force is applied at the center using a knife edge until it breaks, or until the sample is deformed to a specific point if assessing texure characteristics.The degree of softness or freshness of bakery products can be evaluated based on their flexure and fracture properties. However, these properties are desirable only in certain products, while for others, the moisture content is a sign of staleness.

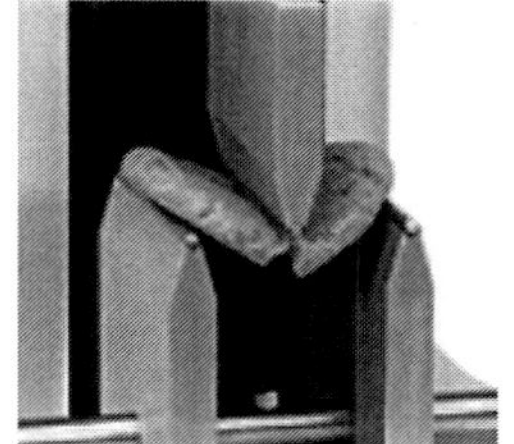

Fig. 7. Three point bend test for biscuits

Tension

Tension testing is used to assess the cohesiveness of components present within them (Figure 8). In this test, a ball probe is pressed through a sample to measure its tensile strength. The freshness of biscuits is indicated by its break resistance and fracture properties. Other properties like extensibility, elasticity and stretching until break of dough, raw materials and gluten can be evaluated through tension testing. In addition to this, the rollability of flour tortillas can be measured by testing the burst strength. The stickiness and adhesion of dough can be evaluated as well.

Fig 8. Tension testing

Specifications and Hygiene Requirements of Biscuits

Biscuits are manufactured from biscuit dough prepared by compounding a mixture consisting primarily of Maida and/or wheat flour, Fat or Bakery Shortening, Sugar and Water. Along with these essential ingredients, some other admixtures like cereals and used products, Oil Seeds Products, Edible Starches, Milk and Milk Products, Spices and Condiments, Food Additives, etc. may be used in combination and suitable proportion for preparation of dough, as per specific requirements. Among these, the synthetic ingredients used shall be of food grade quality conforming to the requirements of PFA Act, 1954 and the rules framed there under.

Biscuits are manufactured in varieties covering sweet, salted, filled and coated biscuits. Based on sensory attributes, the varietal differences can be easily distinguished. Based on this, biscuits have been broadly classified into the following five types:

- Type I – Sweet : They have higher fat and sugar, although their fat content is lower than that of crackers. They are quite crisp and sweet. One example is Parle G.
- Type II - Semi – sweet: They are the likes of Marie — low in sweetness, low in fat and hard in texture
- Type III – Crackers: They are the salted biscuits, crisp and with a high fat content to get the crispness. e.g., 50–50, Monaco etc.
- Type IV - Cookies
- Type V - Speciality Biscuits

Biscuit has nowadays become a food item of mass consumption and has got wide acceptance among all the levels and class of consumers. The Indian Standard specification for biscuits i.e. IS 1011:2002 prescribes both physical as well as chemical tests:

1. Physical requirements (as per Clause 6.1)
2. Chemical requirements (as per Table 1):
   (a) Moisture, percent by mass
   (b) Acid Insoluble Ash (on dry basis), percent by mass
   (c) Acidity of extracted fat (as oleic acid), percent by mass

Besides the above, the standard also prescribes requirements for packaging i.e. containers and materials used for packing biscuits shall conform to the requirements specified in Clause 7.1 of IS 1011: 2002. For packets weighing 150 g and above, the standard specifies that complaint slip should be included in the containers/ printed on the labels. A detailed list of test equipments, apparatus and chemicals required to test for chemical requirements as prescribed in IS 1011:2002. The standard also prescribes hygienic conditions which are required to be followed by the manufacturer within their premises during mixing, processing, handling, packing and storage of finished biscuits, details of which are laid down in IS 5059. The IOs are expected to assess the hygienic conditions being followed by a manufacturer in totality against the requirements of the standard. The standard also prescribes labeling provisions under clause 7.2. Filled or coated biscuits, which are specialty biscuits, are sandwiched with a filling of cream, which has got permitted flavours and permitted food colours among other ingredients. This is reflected in the marking clause which provides for a declaration by the manufacturers that the biscuits contain permitted colours and added flavours. As far as possible, labeling of the product shall be done in accordance with IS 7688 – Code of Practice for labeling of Pre-Packaged Foods. Biscuits are manufactured using the following raw materials in suitable combinations and proportions based on requirements. The probable sources of availability of some major raw materials are given below:

Table Raw material quality checking criteria

Name of the raw material	Source of procurement	Whether or not accompanied by test certificate	Whether tested in -house
Maida/ Wheat flour	Normally procured from open market / flour mill	No; but test certificates are issued by the mills only on request	Yes
Sugar	Open market	No	Yes
Fat / shortening	--do--	No	Yes
Edible salt	--do--	Conforming to IS 253 or IS 7224	—
Edible food colour	--do--	With ISI mark/ certified safe colour	—

In-process Quality Control checks

During the process of manufacture certain quality controls checks are carried out by the manufacturers at the various stages of production. These can be briefly given as follows:

i. Preparation of ingredients – Correct weight of each major ingredient
ii. Creaming of fat, sugar, additives and mixing of dough – Homogeneity, consistency and machinability
iii. Moulding and cutting – Raw weight of the biscuits
iv. Baking – Heat Profile, Baking time
v. Cooling – Weight, Gauge, Colour, Flavour, Texture, Crispiness
vi. Packing – Correct weight of packet, Moisture

Table Raw specification for Biscuits

S. No.	Raw material	Parameter	Standard
1.	Maida	Gluten (%) Sedimentation value (mL) Moisture (%) Ash (%) Acid insoluble ash(%)	7.5–10 18–25 14 (Max) 1(Max) 0.1(Max)
2.	Sugar	Moisture (%) Sugar retention (%) Solubility (%)	0.1(Max) 8 (shell) 4 (cream) 99.9
3	Oil	FFA (%) Peroxide value Moisture (%)	0.1(Max) 1.5 (Max) 0.1(Max)
4.	Sodium meta Bisulphate	pH Purity (%)	4–4.5 90
5.	Ammonium bicarbonate	By mass (%)	99
6.	Starch	Moisture (%) Total ash (%) Ash insoluble acid (%) pH	13(Max) 0.5(Max) 0.1(Max) 4.5–7
7.	Salt	Moisture (%) Temperature(°C) Water insoluble (%) NaCl in water (%) Iodine content (ppm)	1.25 (Max) 105 1(Max) 2.2 15
8.	Condensed Milk	Moisture (%) Milk protein (%) Fat (%)	32(Max) 34 3.9–4.1

Flour quality is highly critical in manufacture of bakery products like biscuits, breads and cakes. Flour with 13–14% moisture, 7.5–10% gluten, 18–25 mL sedimentation value, 1% ash, 0.1 % acid insoluble ash is highly suitable for biscuit making. With respect to sugar, two parameter namely Moisture (1%) and solubility 99.9% were considered best. Peroxide value and free fatty acid value of procured fat should not be more then procured fat should not be more them 1.5 meq/kg and 0.1 respectively. The sedimentation value also decides the utility of the flour in baking industry.

- Soft wheat flour having sedimentation volume less than 20 mL is good for biscuits/ cookies/cakes.
- Medium string wheat flour having sedimentation volume for 20-40 mL is found useful for fermented biscuits
- Hard wheat flour having sedimentation volume more than 40 ml is found for making breads

This apart, the conveyor speed and the temperature of the oven are also monitored closely to take care of any deviations in the process. As the product is sensitive to contamination, the entire process of manufacturing, right from pre-mixing of raw materials to the storage of packed material needs to be carried out under controlled hygienic condition. The requirements given in IS 5059, as given under Annex II, provide detailed guidelines for preventive measures to take care of all types of contaminations. Sources of contaminations may be related to equipment and pipelines used for various processes, product containers, environmental conditions, surroundings, personnel hygiene etc.,

Packaging

General Requirements

1. A utensil or container made of the following materials or metals, when used in the preparation, packaging and storing of food shall be deemed to render it unfit for human consumption:—
   (a) containers which are rusty;
   (b) enameled containers which have become chipped and rusty;
   (c) copper or brass containers which are not properly tinned
   (d) containers made of aluminium not conforming in chemical composition to IS:20 specification for Cast Aluminium & Aluminium Alloy for utensils or IS:21 specification for Wrought Aluminium and Aluminium Alloy for utensils.

Containers made of plastic materials should conform to the following Indian Standards Specification, used as appliances or receptacles for packing or storing whether partly or wholly, food articles namely :—

(i) IS : 10146 (Specification for Polyethylene in contact with foodstuffs);
(ii) IS : 10142 (Specification for Styrene Polymers in contact with foodstuffs);
(iii) IS : 10151 (Specification for Polyvinyl Chloride (PVC), in contact with foodstuffs);
(iv) IS : 10910 (Specification for Polypropylene in contact with foodstuffs);
(v) IS : 11434 (Specification for Ionomer Resins in contact with foodstuffs);
(vi) IS: 11704 Specification for Ethylene Acrylic Acid (EAA) copolymer.
(vii) IS: 12252 - Specification for Poly alkylene terephathalates (PET).
(viii) IS: 12247 - Specification for Nylon 6 Polymer;
(ix) IS: 13601 - Ethylene Vinyl Acetate (EVA);
(x) IS: 13576 - Ethylene Metha Acrylic Acid (EMAA);
(xi) Tin and plastic containers once used, shall not be re-used for packaging

of edible oils and fats; Provided that utensils or containers made of copper though not properly tinned, may be used for the preparation of sugar confectionery or essential oils and mere use of such utensils or containers shall not be deemed to render sugar confectionery or essential oils unfit for human consumption

Nutritional Information

Nutritional Information or nutritional facts per 100 gm or 100ml or per serving of the product shall be given on the label containing the following:

(i) energy value in kcal;

(ii) the amounts of protein, carbohydrate (specify quantity of sugar) and fat in gram (g);

(iii) the amount of any other nutrient for which a nutrition or health claim is made: Provided that where a claim is made regarding the amount or type of fatty acids or the amount of cholesterol, the amount of saturated fatty acids, monounsaturated fatty acids and polyunsaturated fatty acids in gram (g) and cholesterol in milligram (mg) shall be declared, and the amount of trans fatty acid in gram (g) shall be declared in addition to the other requirement stipulated above;

(iv) Wherever, numerical information on vitamins and minerals is declared, it shall be expressed in metric units;

(v) Where the nutrition declaration is made per serving, the amount in gram (g) or milliliter (ml) shall be included for reference beside the serving measure;

Declaration regarding Food Additives-

(i) For food additives falling in the respective classes and appearing in lists of food additives permitted for use in foods generally, the following class titles shall be used together with the specific names or recognized international numerical identifications: Acidity Regulator, Acids, Anticaking Agent, Antifoaming Agent, Antioxidant, Bulking Agent, Colour, Colour Retention Agent, Emulsifier, Emulsifying Salt, Firming Agent, Flour Treatment Agent, Flavour Enhancer, Foaming Agent, Gelling Agent, Glazing Agent, Humectant, Preservative, Propellant, Raising Agent, Stabilizer, Sweetener, Thickener:

(ii) Addition of colours and/or Flavours—

(a) Extraneous addition of colouring matter to be mentioned on the label – Where an extraneous colouring matter has been added to any article of food, there shall be displayed one of the following statements in capital letters, just beneath the list of the ingredients on the label attached to any package of food so coloured, namely: Contains Permitted Natural Colour (s) or Contains Permitted Synthetic Food Colours(s) or Contains Permitted Natural and Synthetic Foods Colour (s) Provided that where such a statement is displayed along with the name or INS no of the food colour, the colour used in the product need not be mentioned in the list of ingredients.

(b) Extraneous addition of flavouring agents to be mentioned on the label. Where an extraneous flavouring agent has been added to any article of food, there shall be written just beneath the list of ingredients on the label attached to any package of food so flavoured, a statement in capital letters as below : Contains added flavour (specify type of flavouring agent as per Regulation 3.1.10(1) of Food Safety and Standards (Food product standards and food additive) Regulation, 2011

(c) In case both colour and flavour are used in the product, one of the following combined statements in capital letters shall be displayed, just beneath the list of ingredients on the label attached to any package of food so coloured and flavoured, namely :— Contains Permitted Natural Colours (s) and added flavour (s) or Contains Permitted Synthetic Food Colours(s) and Added Flavour(s) or Contains Permitted Natural and Synthetic Food Colour(s) and Added Flavour(s) Provided that in case of artificial flavouring substances, the label shall declare the common name of the flavours, but in case of the natural flavouring substances or nature identical flavouring substances, the class name of flavours shall be mentioned on the label and it shall comply with the requirement of label declaration as specified under the regulation 2.2.2 (5) (ii) Note: — When statement regarding addition of colours and/or flavours is displayed on the label in accordance with regulation 2.2.2(5) (ii) and regulation 3.2.1 of Food Safety and Standards (Food Product Standards and Food Additive) Regulation, 2011, addition of such colours and/or flavours need not be mentioned in the list of ingredients. Also, in addition to above statement, the common name or Version –I (08.05.2017) class name of the flavour shall also be mentioned on label. Provided further that when combined declaration of colours and flavours are given, the International Numerical Identification number of colours used shall also be indicated either under the list of ingredients or along with the declaration. Provided also further that every package of synthetic food colours preparation and mixture shall bear a label upon which is printed a declaration giving the percentage of total dye content.

References

Cauvian, SP. Quality Measures in Baking. Cereal Foods World. May-June. 2013 Vol.58 No.3

Dobraszczyk BJ, Dendy DAV (2001). Cereal and Cereal Products: Chemistry and Technology. Aspen Publisher, Inc.

Manual for Small Scale Bakery Units, FSSAI

9

Confectionery and Chocolate Ingredients

Sugar confectionery and chocolate includes candies (*sweets* in British English), candied nuts, chocolates, chewing gum, bubble gum, pastillage, and other confections that are made primarily of sugar. In some cases, **chocolate confections** (confections made of chocolate) are treated as a separate category, as are sugar-free versions of sugar confections. The words *candy* (US and Canada), *sweets* (UK and Ireland), and *lollies* (Australia and New Zealand) are common words for the most common varieties of sugar confectionery. Confectionery depends on numerous ingredients which have been discussed in detailed in this chapter

Sugar

Sugar is in fact a generic name referring to a host of carbohydrates, but it has become common usuage for it to refer to one particular substance- sucrose. Sucrose is produced in vast quantities through out the world and is the basic ingredient for classical sugar confectionery. Sugar occurs very widely in the vegetable world, in the roots and stem of grasses and root vegetables and in the sap of many trees. Commercially, however, it is extracted from sugarcane, which is grown in tropical areas, and from sugar beet, which is grown in temperature climates.

Some important types of sugar are as follows:

1. Granulated
2. Caster
3. Icing
4. Liquid sugars
5. Brown sugars
6. Molasses
7. Micro Crystalline sugars

1. Granulate: it is again differentiated into different grades

(a) Mineral Water

This is the purest grade of sugar which is commercially available. It has a lower color and ash content than granulated and was originally produced for the manufacture of mineral water. However, it has found use in sugar confectionery for specific purposes, eg., Wet crystallization.

(b) Granulated

A white sugar which is sold industrially and domestically and which constitutes a very high proportion of total production.

(c) Industrial Granulated

This sugar has very slight off white-colour and is used where a white sugar is unnecessary. Eg-in the manufacture of Toffee, Fudge, Chocolate etc.,

(d) Cubes

These are usually produced now a days by moistening granulated sugar with about 1% water pressing into cubes and drying.

(e) Nibs

These are agglomerates of granulated sugar crystals made by dampening the sugar, thoroughly drying and breaking up the resulting hard mass. The product is sieved to various sizes.

2. Castor

A white sugar of small crystal size, for domestic and industrial use. It can be either boiled in the vaccum pans or milled from granulated and these five have the same composition as granulated.

(3) Icing

Icing is produced by milling granulated sugar, the best qualities being double milled. Often an insoluble anti-caking agent is added.

(4) Liquid Sugars

There are advantages and disadvantages in using liquid sugars, for example ease of handling and being already dissolved against the extra water to transport and to evaporate. When total solids are under 75%, precautions must be taken to avoid microbial spoilage. Liquid sugars come in many forms. Mineral water or granulated sugar can be dissolved in distilled to provide the highest quality. For most purposes in the confectionery industry, the decolorized liquor from the refining process is supplied, instead of evaporating it to produce granulated sugar. Relatively impure mother liquor (i.e., those which contain too much colour, ash etc to produce white sugar) can be sold as such, when these small amount of impurities are not of prime importance and show savings in cost.

Still lower grade sugar can be used to provide colour and flavour. All or part of sugar may be inverted and the total solids may be between 66-84%.

(5) Brown Sugars

The first is, in effect, raw sugars produced by the sugar factories. These can have good flavour but some may suffer from variation in quality, particularly regarding hygiene and foreign matter.The second type is produced in sugar refineries. Originally they were _boiled' sugars, that is, impure sugar solutions were evaporated until brown sugar of various qualities were produced. However the process is very slow and now a days brown sugar produced by combining mixtures of impure syrups with white sugars of the appropriate crystals size. This process also gives more uniform products.

(6) Molasses

A cane refinery will produce 600-800 tonnes of molasses per week, from which no more sugar can be extracted. Some is used for human consumption, but the bulk goes to cattle food and to the fermentation industries. Eg., alcohol and citric acid.

Treacle is clarified molasses and can be mixed with higher purity syrups to mellow the taste. Within limits, its composition can be altered to simplify product formulations.

(7) Micro Crystalline Sugars

Sugar syrup is evaporated to around 95% solids and then subjected to intense shear. The sugar crystallizes instantly as very fine crystals (5-20mm), the dried, milled and sieved final products being agglomerates of these crystals. Sugars produced in this way have special properties- For example-whites dissolves very rapidly and browns are free flowing.

Alternative bulk Sweeteners

Alternative bulk sweeteners can broadly be divided into two categories:

1. Sugar
2. Sugar Alcohols

Alternative sugars are generally used to replace a proportion of the sucrose in confectionery products in order to modify the sweetness and textural properties. Sugar alcohols (also known as polyols) are generally used to replace all of the conventional carbohydrates in the manufacture of non cariogenic , diabetic or dietetic confections.

I. Alternative Sugars

1. Glucose

The mono sacturide glucose occurs widely in nature where it is found together with fructose in most fruits and in honey. It is commonly called as dextrose.

Glucose has a lower sweeteners, lower solubility and lower viscosity than sucrose. It is however a better humectants and provides better preservative properties owing to its lower water activity. It also has a noticeable cooling effect arising from negative heat of solution.

2. Fructose

Fructose also known as laevulose or fruit sugar is another monosaccharides found in fruits and honey. Fructose may be manufactured from sucrose by isolation from invert sugar or from starch by isolation from high fructose glucose syrup. Fructose is considerably more hygroscopic and has better humectant properties than either sucrose or glucose. It is also more reactive, being a ketose rather than an aldose sugar. Hence fructose has an even greater tendency to browning than glucose.

3. Lactose

Lactose also known as milk sugar, is a disaccharide molecule comprising glucose and galactose. The main commercial source is whey from which it is extracted by either crystallization or precipitation. Lactose is usually crystallized as the α monohydrate, which melts at 202 °C, although some beta lactose anhydride (Melting Point - 252 °C) is also produced for special applications. Maximum sucrose replacement levels of 5–35% have been suggested for various confectionery applications. In non grained confections the lactose content should not exceed 10%.

II. Sugar Alcohols & Its Confectionery Applications

1. Sorbitol

Sorbitol is primarily used in manufacture of diabetic and sugar free confections. The most important applications includes chewing gum, compressed mints, high boiling, gums, pastilles and chocolates. In chewing gum sorbitol is typically used together with maltitol syrup which provides the liquid phase and saccharin or aspartame which are needed to boost the sweetness. A proportion of mannitol may be included in order to inhibit crystallization if desired chewing gum dragees may be hard coated with sorbitol. Sorbitol cannot be used to manufacture high boiling by conventional means due to its low viscocity.

2. Xylitol

Xylitol is currently used in a variety sugar free and diabetic confectionery products. Xylitol can be used in the hard panning of chewing dragees or of other confectionery centers. Xylitol can act as the sole sweetner in recrystallized hard candies. Fondant is another application for which xylitol is well suited.

Xylitol solutions are lower in both viscocity and water activity then equivalent concentrations of other polyols but do not have particularly good humectants properties.

3. Maltitol

Maltitol syrups have been used in a wide variety of applications either alone in combination with other polyols. Their function is sugarless chewing gums and in high boilings which can be manufactured simply by adding acid, color and flavor to a boiled maltitol syrup. Caramels and chews can be prepared from maltitol syrup.

4. Isomalt

Isomalt can be used in a variety of sugar free confectionery products including high boiling, compressed tablets, marzipan, chews, liquorice and chocolate. In other applications isomaltase best used in combination with other polyols which are needed to inhibit crystallization as well as to increase the sweetners. Compressed tablets also benefits from the low hygroscopicity of isomalt. A wet granulation stage is required in order to improve the compressibility.

5. Polydextrose

Polydextrose can be used in the manufacture of high boiling since it forms a stable glass structure. It also reduces the viscosity thus improving handling properties. In the context it has been successfully combined with xylitol in reduced calorie chews with sorbitol or xylitol in gelatin jellies and with isomalt in fondant. In addition to its application in sugarfree products polydextrose is used in combination with fructose for diabetic lines or with sucrose in standard reduced calorie lines.

Gelling and Whipping agents

Gelling and Whipping agents are used to provide a wide range of textures in the sugar confectionery industry.

1. Agar agar E406:

Agar agar is dried hydrophilic, colloidal polysaccharide from red seaweeds and related marine species. It is available as white to pale yellow agglutinated strips or in flate or powder form. It may have a slightly characteristic odour and mucilaginous flavour. It is soluble in boiling water and insoluble in cold water and most organic solvents. It has a molecular weight of over 20,000. Agar is extracted from a wide range of seaweed varieties which grow in many areas of the world. The main suppliers are Japan, New Zealand, Denmark, Australia, South Africa and Spain. The gel strength varies according to the source and checks should be carried out on each delivery to determine gel strength.

Normally agar is dissolved in 30-50 times its weight of water, usually premixed with about 10 times weight of sugar to prevent lumping. Very high viscosities are achieved with concentration upto 10%. Agar provides good gel strenth. It forms a firm gel at concentrations as low as 1% and is usually used in confectionery at the level of 1-1.5% of sugar glucose agar recipe. Agar is not absorbed by the body during digestion and can therefore be used in low-calorie confections. Agar does not carry flavour well, and as a result of this and its sensitivity to acid and particular types of texture it is being replaced by pectins or modified starches.

2. Alginate E401

Alginate were first is isolated by Stamford by alkaline extraction from brown algae, a process used for iodine production. Commercial extraction is from seaweeds such as *Laminaria digitata, Ascophyllum nodosum and Fucus serratus*. Each of the seaweeds provides a differing proportion of the main attribute of alginate. It is a white to yellow granular powder, colloidal, insoluble in water, acids and organic solvents. Alginates are comprised of mannuronic and guluroni acids. These can link to form homogeneous segments in which guluronic acid binds to guluronic acid and mannuronic acid binds to mannuronic acid.

3. Carrageenans

The name carrageenans is derived from the country of Carraghen on the south coast of Ireland, where Irish moss was used in foods and medicines more than 600 years ago. Red seaweeds were used because of their unique property in gelling milk when they are heated together. The carrageenan coagulates into fibres, leaving impurities in the solution. This product is pressed and washed again with alcohol to complete its dehydration. It is then dried under vacuum, milled and sieved to the exact particle size.

The gelation of *k (kappa)* and i(iota) carrageenans is induced by the association of chains through double helices.

4. Gelatin

Gelatin does not exist naturally but is produced by the partial hydrolysis of collagen in the raw material substrate. Collagen is a structural component in animal tissues, present in skin, bone and connective tissue. The raw materials are sourced from slaughterhouses, meat-packing plants or tanneries. The products from tanneries have already been salted or limed for preservation. Collagen is made up of films and fibrils. Industrial modification of collagen to produce gelatin is by stepwise destruction of the organized structure to obtain the soluble derivative gelatin. The extraction of gelatin from the raw material is initiated by either liming or acidulation, which disrupts the molecular linkages within the collagen. Gelatin is then extracted by hot-water hydrolysis. This is carried out as a batch operation. Several extracts are produced with a concentration of 5–10% gelatin. A typical analysis of a gelatin would be Moisture 14%, Protein 84% and Ash 2%

Gelatin picks up water in a moist atmosphere and should be stored in a cool dry store. At about 16% moisture mould growth is possible. Gelatin solutions form an ideal medium for bacterial growth. Hygienic procedures must be implemented when using this product in solution and equipment must be thoroughly cleaned.

5. Pectin

Pectic substances are matrix components in the cell walls of higher plants. The compounds are insoluble in aqueous solution and are referred to as protopectins. Protein consists mainly of the partly methylated esters of polygalacturonic acid and their ammonium, sodium, potassium or calcium salts. The molecular weight is between 20,000 and 100,000. The protopectin is hydrolysed using acid in hot aqueous solution. The aqueous extract contains soluble products such as neutral polysaccharides, gums and others. LM pectins are defined as having a degree of methoxylation of less than 50% i.e., less than 50% of the functional groups on the molecule are methoxylated. The grade strength of pectin is defined as the number of grams of sugar with which one gram of pectin will produce a gel of standard firmness, when tested under standard conditions of acidity and soluble solids content.

During cooking the product mixture is normally buffered to maintain the pH within controlled limits. Pectins can be purchased pre-buffered or the manufacturers can add citrates, tartrates, etc, to act as the buffering agent.

6. Xanthan gum E415

A polysaccharide gum produced by *Xanthomonas campestris*. It occurs as a cream-coloured powder, soluble in water, insoluble in alcohol. Xanthan gum is a secondary metabolite of Xanthomonas campestris produced during the commercial aerobic fermentation of carbohydrates. Fermentation is carried out in a batch process. The gum is recovered from the broth by the addition of propan-2-ol. The precipitate obtained is washed and pressed to remove residual alcohol. Xanthan gum is a mixed polysaccharide with a molecular weight of approximately 2.5 million. The monomer units are D-glucose, D-mannose and D-glucoronic acid. Xanthan gum is readily dissolved in hot or cold water to produce an opaque solution of relatively high viscosity. This solution exhibits pseudoplastic flavour characteristics, i.e the viscosity of the solution decreases rapidly when shear is applied. As the shear rate decreases there is an immediate return to the high original viscosity. This charecteristic makes Xanthan an excellent suspending agent at low concentration. Xanthan forms a thermo reversible cohesive gel system with locust bean gum.

Confectionery Fats

Vegetable fats generally are used in great quantities in the production of all kinds of confectionery, such as caramels, fudge, nougat, truffles, and pastes for wafer and cookie fillings. The only animal fat normally used in these products is butter.

1. Soya Oil

The soya bean had, its origin in eastern Asia but enormous expansion of its cultivation has occurred in the United States during this century.

2. Cottonseed Oil

It was long after the plant was cultivated, for cotton that the seed became an important source of vegetable oil. The mature plant produces fluffy white seeds called cotton bolls with the fiber adhering. The seeds are oval, and yield from 15–25 percent of oil.It is a plant of tropical or subtropical regions.

3. Sunflower Seed Oil

The plant is very tall (5–8 ft), althougp there are dwarf varieties. The flower has a dark-brown center with yellow petals. It is native to, Central America but is now cultivated in many parts of the world.

4. Sesame Seed Oil

Sesame seed originated in china and India. Today it is also grown extensively in Africa and Mexico. It is a crop that grows in poor soil and is easily cultivated. The seed contains about 50 percent oil, which has uses similar to those of olive oil.

5. Rapeseed Oil

Rapeseed can be grown in colder climates and in recent years countries like Sweden, Denmark, Poland, and Canada have increased production considerably, thereby reducing consumption of imported tropical oils. The plant is of the Brassica (cabbage) family and fields when in flower are brilliant yellow. Rapeseed contains 35–40 percent oil. Rape oil from original seed contained a high proportion of erucic acid that has been shown to be dietetically undesirable. New genetic varieties of seed have been developed giving oil of low erucic acid content.

6. Olive Oil

The olive tree has been a source of edible oil for many centuries and olive oil, although perhaps now not so important commercially, is the highest-quality vegetable oil and greatly prized for table use. The tree grows in the countries around the mediterranean, with Spain and Italy the main producers. The fruit contains about 15 percent of oil.

7. Corn Oil

This has been an important edible oil of recent years, produced as a by-product of the vast starch, glucose syrup, and dextrose industry. Corn oil is pale yellow in color, liquid at normal temperatures but deposits a small amount of stearine at lower temperatures. The oil is almost entirely in the germ, which is separated in the early stages of wet milling and starch extraction. Corn is a major crop in the united states and as a result of intensive scientific development there is an ever-increasing supply of derived products-not only in the food industry but also such materials as adhesives and paper.

Food Colours and Flavors

1. Colours

When synthetic colours were first added to foods the dyes used were merely batches of the sort of dye used in the textile industry. The use of colours in foods is strictly regulated. Governments around the world have lists of permitted colours. Unfortunately, the lists differ throughout the world. It might be thought that some scientific consistency could be achieved but this is not the case. Manufacturers who produce products for the international export markets are reduced to leaving out all the colours as a way of making the product universally acceptable. Early fruit flavoured products were probably flavoured with jam and did not have a particularly strong flavour.

Technical requirements of colours in Bakery products

To be used successfully in bakery products a food colour needs the following attributes as well as complying with the appropriate legislation:

- It should be stable to heat and light, stable to reducing sugars, and raw materials. Resistance to sulfur dioxide is useful.
- Most colours used in bakery products are water soluble. This is simply convenience; some flour confectionery products contain very little fat any way.

1. Synthetic colours

Synthetic colours are available for almost all possible shades. Intermediate shades can be produced by blending colours. In general, Synthetic colours are much more stable than natural coloours to light, heat and extremes of pH.

Synthetic colours can be supplied as soluble powders, prepared solutions, easily dispersed granules, pastes or gelatine sticks. Blocks of colour in vegetable fat are available for use in fat-based products. The attraction of soluble powders is that they are the least expensive and can be made up as required for use. The other forms have the advantage that they are at a concentration that is ready to use.

The disadvantage is usually financial Synthetic colours are normally so intense that they must be considerably diluted for them to be readily measured and dispersed into the product. Colour solutions made up in the factory have to be prepared not more than twenty four hours before use to avoid mould spoilage. The pre-prepared colour solutions will contain a permitted preservative or will be made up in glycerine, propylene glycol or propan-2-ol. These non-aqueous solvents inhibit mould growth lists some synthetic colours.

2. Natural colours

There is a belief that natural products are inherently safer and healthier than man-made ones. This belief is lacking in intellectual rigour. Of the most toxic substances known to man most are natural, eg., aflatoxin, a mould metabolite, and ricin, found in castor oil beans. However, the presence of natural colours is a marketing advantage and so they are used. Natural colours in general are less heat stable, less light stable and give a less intense and less pure colour than Synthetic colours.

Natural colours have been used in the form of impure extracts rather than pure products. In this form higher doses are needed than with synthetic products. When purified, some natural pigments are more intense in colour and can be used in lower doses than Synthetic colours. One other problem with natural colours is that the range of colours available is restricted. Several sources of natural colours are given in the following subsections.

(a) Caramel (E150)

Caramel in this context means a brown colour that is produced either traditionally by heating sugar or as a very intense product that is made by heating carbohydrate, usually glucose syrup, with ammonia. Caramel colour is the product of the Maillard reaction, i.e the reaction of a reducing sugar with an amino group. Chemically the colour is a melanoidin. These substances are extremely stable chemically and can be used in any type of confectionery.

(b) Chlorophyll

This is the green pigment that is responsible for photosynthesis. It is widely distributed in nature, sources are green leaves, grass, alfalfa and nettles. The extract that is used is a mixture of chlorophyll with lutein and other carotenoids. This product gives an olive green colour. Chlorophyll is most stable in neutral or alkaline conditions but has a limited stability to heat and light. Chlorophyll preparations are available for colouring boiled sweets. This is made from chlorophyll. It is more blue than natural chlorophyll. The chemical modification makes it much more stable to heat and light. It is a more useful material than natural chlorophyll.

(c) Cochineal (E120)

Cochineal is a traditional natural colour. It is made from a Mexican beetle. The only problems with cochineal, apart from expense, is that it is not kosher and it is not animal free. Cochineal is not kosher not because it is made from an insect but because the insect is not itself kosher.

(d) Riboflavin

This is vitamin B2. Riboflavin can be extracted from yeast but is normally encountered as a nature identical substance. Unfortunately, riboflavin has an intensely bitter taste. The colour produced is an orange yellow. It is stable to acid but is unstable in water. Riboflavin is sometimes used for panned goods.

(e) Riboflavin-5-phosphate (E 101 a)

This material is both less bitter and more water stable than riboflavin. It is normally only encountered as a pure synthetic substance. Like riboflavin it is used on panned products.

(f) Carbon Black

This is carbonized vegetable matter, ie, very finely divided char coal. Inevitably it is the most light fast of all colours. Obviously, it is only available as a solid. A common use is in liquorice products.

(g) Curcumin (E100)

Curcumin is obtained from the spice turmeric, which comes from the plant curcuma longa, of the ginger family. Curcumin is obtained by extraction from the plant to give a deodorized product. Curcumin is a bright yellow pigment that is oil soluble. It is sometimes produced in a water dispersible form. The colour of curcumin varies with the pH of the medium. Under acid conditions a bright yellow is obtained but under alkaline conditions a reddish brown hue is obtained. This colour shift occurs because curcumin undergoes keto-enol tautomerism. The most serious problem with curcumin is instability to light. One recommendation is that curcumin should not be used in products that are exposed to light unless the moisture content is very low. A confectionery product that fits this description is boiled sweets. The heat stability of curcumin is sufficiently good that it can withstand 140 °C for 15 min. in a boiled sweet mass. The other stability problem with curcumin is sulfur dioxide. If the sulfur dioxide level is above 100 ppm then the colour will fade. Within the restrictions outlined curcumin is a successful natural colour.

(h) Carotenoids

The carotenoids are a wide range of substances. They are extremely abundant in nature. Some 400 carotenoids have been identified to date. They are found in fruits, vegetables, eggs, poultry, shell fish and spices. Orange juice and peel contain 24 different carotenoids.

Legally, carotenoids are divided between two E numbers. E 160 covers the carotenoid hydrocarbons b-carotene, lycopene and paprika as well as the apo-carotenoids, e.g., bixin. E 161 covers the xanthophylls and the carotenoids lutein, astaxanthin and canthaxanthin. Most carotenoids are fat soluble, although preparations that allow them to be dispersed in water are made. The colours available from acrotenoids vary between pale yellow and red. Chemically carotenoids have conjugated double bonds that render them liable to oxidation. This tendency to oxidation can be diminished by adding antioxidants to the product. In the sort of product where natural colours are used suitable antioxidants would be tocopherols or ascorbic acid.

Chemically antioxidants such as butylated hydroxytoluene might be suitable technically but would not fit the image of an all natural product. Ascorbic acid could be declared as vitamin C rather than as an antioxidant. Oxidation can be started by exposure to light and so this is best avoided. Carotenoids are generally stable to heat.

The levels required can be as low as 10 ppm. b-carotene is available as a nature identical form.

(i) Crocin

Crocin is found in saffron and in gardenias. Extracting crocin from saffron is not economically viable. Saffron is obtained from Crocus sativus. Seventy thousand plants are needed to produce 500g of saffron, which would contain 70g of crocin. The commercial source of crocin is the gardenia bush. The town of Saffron Walden in Essex, UK, takes its name because saffron used to be produced there. Chemically, crocin is the digentiobioside of crocetin. It is one of the few water-soluble carotenoids to produce a bright yellow shade in water. Unfortunately, crocin is bleached by sulfur dioxide levels above 50 ppm. the heat stability of crocin is good enough to use it in boiled sweets.

(j) Annatto [E160 (b)]

Annatto is classified as E160(b). it is extracted from the seeds of a tree (Bixa orellana), which grows in America, India and East Africa. The extract is a mixture of two pigments, bixin and nor –bixin. Bixin is oil soluble while nor-bixin is water soluble. Both bixin and nor-bixin produce orange solutions. Bixin produces an orange solution in oily media while nor-bixin produce orange aqueous solution. Obviously, bixin is the product of choice for high fat systems while nor-bixin is used in aqueous systems. Nor-boxin is one of the two water-soluble carotenes. Nor-bixin is damaged by sulfur dioxide if the concentration exceeds 100 ppm. Acidic conditions or divalent cations, particularly calcium, can cause nor-bixin to precipitate. These problems are tackled by producing nor-bixin preparations with buffers and sequestrants. Nor-bixin is relatively stable to heat. The most severe conditions will either isomerise the pigment or shorten the chain. Either of these changes will make the pigment more yellow. Nor-bixin can associate with protein, which stabilizes the nor-bixin. The other effect of this association is to redden the colour.

(k) Lutein [E161 (b)]

Lutein is one of the four most common carotenoids found in nature. The EU classifies it as E161(b). chemically, lutein is a xanthophylls and is similar to b-carotene. Although lutein occurs in all green leafy vegetation, egg yolks and in some flowers the commercial sources are the petals of the Aztec marigold and to a lesser extent, alfalfa. Purified alfalfa gives a clean, bright lemon yellow shade. Lutein is more stable to oxidation than the other carotenoids. It is also resistant to the action of sulfur dioxide. Lutein is oil soluble and is most effective dissolved in oil. Aqueous dispersible preparations based on lutein are available.

(l) Betalaines

The main pigment in the concentrated colour beet red is betanin. This is classified as E 162 by the EU. The pure pigment is obtained by aqueous extraction of the red table beet. Approximately 80% of the pigment present in beetroot is betanin. In an aqueous solution betanin gives a bright bluish red. The pure pigment is so intensely coloured that dose levels of a few parts per million are satisfactory. The problems with betanin relate to stability. Betanin is extremely sensitive to prolonged heat treatment. Short spells such as ultrahigh temperature (UHT) are tolerated. The conditions that make betanin unstable are oxygen, sulfur dioxide and high water activity. As confectionery is a low water activity system without sulfur dioxide or oxygen, betanin can be used.

(m) Anthocyanins

Anthocyanins are water soluble and are responsible for the colour of most red fruits and berries. Some 200 individual anthocyanins have been identified. It has been estimated that consumption of anthocyanins is an average of 200mg per day. This is several times greater than the average consumption of colouring material. There are claims made that consuming anthocyanins has health benefits. Chemically anthocyanins are glycosides of anthocyanins and are based on a 2-phenylbenzopyrilium structure. The properties of the anthocyanins depend on the anthocyanidins from which they originate. Anthocyanins are extracted commercially using either acidified water or alcohol. The extract is then vacuum evaporated to produce a commercial colour concentrate.

The raw materials can be black currants, hibiscus, elderberry, red cabbage or black grape skins. The most commonly used commercially are black grape skins, which can be obtained as a by-product. Anthocyanins usually give a purple red colour. Anthocyanins are water soluble and amphoteric. There are four major pH dependent forms, the most important being the red flavylium cation and the blue quinodial base. At pH upto 3.8 commercial anthocyanin colours are ruby red; as the pH becomes less acid the colour shifts to blue. The colour also becomes less intense and the anthocyanin becomes less stable. The usual recommendation is that anthocyanins should only be used where the pH of the product is below. As these colours would be considered for use in fruit flavoured confectionery this is not too much of a problem. Anthocyanins are sufficiently heat resistant that they do not have a problem in confectionery. Colour loss and browning would only be a problem if the product was held at elevated temperatures for a long while. Sulfur dioxide can bleach anthocyanins- the monomeric anthocyanins the most susceptible. Anthocyanins that are polymeric or condensed with other flavonoids are more resistant. The reaction with sulfur dioxide is reversible. Anthocyanins can form complexes with metal ions such as tin, iron and aluminium. The formation of a complex as expected alters the colour usually from red to blue. Complex formation can be minimized by adding a chelating agent such as citrate ions. Another problem with anthocyanins is the formation of complexes with proteins. This can lead to precipitation in extreme cases. This problem is normally minimized by careful selection of the anthocyanin.

(n) Beta-Carotene [E160(a)]

The natural sources that are exploited commercially for b-carotene are carrots and algae. The EU classifies b-carotene as E160(a). b-carotene is an oil-soluble pigment, although forms that can be dispersed in water are available. The colour obtained varies between yellow and orange, depending on concentration. b-carotene is stable to heat, sulfur dioxide and pH changes. It is, however, sensitive to oxidation, particularly when exposed to light. b-carotene is successfully used to colour boiled sweets and other confectionery products.

Flavours

Flavours are complex substances that can conveniently be divided into three groups: natural, nature identical and synthetic.

1. Natural flavours

These can be the natural material itself; one example would be pieces of vanilla pod or an extract, eg, vanilla extract. Extracts can be prepared in several ways. One is to distil or to steam distil the material of interest. Another is to extract the raw material with a solvent, eg, ethyl alcohol. Alternately, some materials are extracted by coating the leaves of a plant with cocoa butter and allowing the material of interest to migrate into the cocoa butter.

These techniques are also used in preparing perfumery ingredients, indeed materials like orange oil are used in both flavours and perfumes. In practice some natural flavours work very well; any problems are financial rather than technical. Examples of satisfactory when all natural. Notably, citrus oils are prepared from the skin rather than the fruit. The view exists that natural products are inherently safer and healthier than synthetic materials. Curiously, any new synthetic ingredient has to be most rigorously tested before it is allowed in foods. Natural products, provided their use is traditional, are normally allowed without testing. There is a legal distinction applied between an ingredient and an additive. In the UK, additives generally need approval while natural ingredients, provided their use is traditional, do not. Periodically, some natural substance is tested and found to have some previously unknown potential risk.

2. Natural Identical Flavourings

These are materials that are synthetic but are the same compound as is present in a natural flavouring material. From time to time it emerges that one substance produces a given flavour. Most chemists know that benzaldehyde has a smell of almonds. Some chemists know that hydrogen cyanide smells of bitter almonds. If a natural flavouring can be represented by a single substance and that substance can be synthesized then the flavour is likely to be available as a nature identical flavour. Vanilla flavour is a good example. Vanilla flavour can be all natural and derived from vanilla pods or nature identical or artificial. The nature identical product would be based on vanillin, which is in vanilla pods and has a flavour of vanilla. An artificial vanilla flavour would be ethyl vanillin, which is not present in vanilla pods but has a flavour two and a half times stronger on a weight basis than vanillin. The claim nature identical does not seem to be much appreciated in the English speaking countries. In some other countries it is an important claim for marketing purposes.

The qualification for nature identical varies between jurisdictions. In the

EU, ethyl acetate made from fermented ethyl alcohol and fermented acetic acid is nature identical. In the USA, provided that the ethyl alcohol and acetic acid are natural, i.e, produced by fermentation, the ethyl acetate would be natural. Practical flavours often contain a mixture of substances, some natural, some nature identical some synthetic. UK law classifies a flavour that contains any nature identical components as nature identical even though the rest of the flavour is natural. Similarly, the presence of any artificial components renders the flavour artificial.

The case for Nature Identical flavours. Although not much appreciated in English speaking countries, nature identical claims are more popular in German-speaking countries. Presumably the advantage of a nature identical substance is assumed because it is thought to be inherently safe. This is a paradox since synthetic substances are normally tested for safety much more exhaustively than natural ones. Nature identical flavours do have the advantage over natural products that the price or quality is not affected by adverse harvests.

3. Synthetic Flavours

These are flavours that are produced synthetically but are not present in a natural flavouring material. The chemistry of flavours is a complex topic that has been the subject of many books. Synthetic flavours are made from a mix of flavouring substances that have been found to produce a given flavour note. Those who develop flavours are referred to as flavourists. Flavourists take the musical analogy of notes further by referring to the top notes and the bottom notes of a flavour. Flavour research is driven by a need to find compounds that produce desirable flavours. In some cases the improvement that is sought over the natural substance is not flavour intensity or cheapness but chemical stability.

One view of the way that flavours work is that they interact with certain receptors in the nose. Any other compound that has the same shape will work as well. A typical synthetic flavour is a very complex mixture of substances. The mixture used will have been chosen to give the desired properties in the system of choice. Compounding flavours is a mixture of chemistry and sensory skill.

Modified Starches in Candy and Confectionery Industry

Native starches are structurally too weak and limited functions for application in pharmaceutical, food and non-food technologies due to its inherent weakness of hydration, swelling and structural organization. Unprocessed starches produce weak-bodied, cohesive rubbery pastes when cooked and undesirable gels when the pastes are cooled. To enhance viscosity, texture, stability among many desired functional properties desired, starch and their derivatives are modified by chemical, physical, and enzymatic methods. Modifications are necessary to create a range of functionality.

Starch modification can be introduced by altering the structure and affecting the structure including the hydrogen bonding in a controlled manner to enhance and extend their application in industrial prospective. This modification includes esterification, etherification, cross linking, acid hydrolysis, enzymatic hydrolysis heat treatment and grafting of starch. Modified starches can be found applicable practices in food industry and non-food industry.

Various types of modified starches for wide applications in many industries

1. Pregelatinized starch

It is the simplest starch modification, prepared by cooking. It maintains starch integrity while providing cold water thickening which is a process that breaks down the intermolecular bonds of starch molecules in the presence of water and heat, allowing the hydrogen bonding sites (the hydroxyl hydrogen and oxygen) to engage more water.

2. Cross-linked starch

Cross linking is the most important modified form that used in the food industry. It involves replacement of hydrogen bond present between starch chains by stronger, permanent covalent bonds. Distarch phosphate or, adipate are commonly used in cross-linked starch. Cross-linked starches offer acid, heat and shear stability over the native starch. Food with this type of starch processing tends to have longer shelf life.

3. Oxidized starch

The processing includes reaction with oxidizing agent such as sodium hypochlorite or peroxide. This type of starch is mainly used as surface sizing agent or coating binder and available in different viscosity grade. Oxidized starches have shorter chain lengths than native starches. It improves whiteness and reduces microbiological content. Oxidized starches are the best thickener for applications requiring gels of low rigidity. This improves adhesion in batters and breading.

4. Cationic starch

Cationic starches are produced by reacting native starches with tertiary or, quaternary amines, using wet or dry production processes. They are mainly used in paper forming process. Cationic starch represents high performance starch derivatives widely used by paper manufactures to increase strength and retention. Cationic starches carry a formal positive charge over the entire pH range creating their affinity towards negatively charged substrates, such as cellulose, pulp and some synthetic fibres, aqueous suspensions of minerals and slimes and biologically active macromolecules. Cationic starch is also added at the beater to improve drainage on the wire, better sheet formation, and enhancement of the sizing efficiency of an alum-rosin size.

5. Anionic starch

Anionic starches are prepared by reaction with phosphoric acid and alkali metal phosphates or by making derivative with carboxymethyl group.

6. Thinned starch

These are produced through depolymerisation reaction by hydrochloric acid or other acids. Unmodified starches are treated with a mineral acid at temperature lower than gelatinization and results in partially hydrolyzed starch molecules. This cleaves the chain length and lower viscosity. It increases the tendency to retro gradation. The lower viscosity permits higher concentrations to be used forming rigid gels in gums pastilles and jellies. In these applications, increased set-back leading to the formation of strong gels gives these starches significant advantages over native starches. Extended applications in food industry are found by acid-thinned starch in conjunction with esterification and etherification reaction.

7. Acetylated starch

Acetylated starch (E1420) esterification with acetic anhydride Starch after treatment with acetic anhydride produces starch esters which are useful in biodegradable applications. In particular, high starch acetates provide thermo plasticity, hydrophobicity and compatibility with other additives. The result of this treatment is a stability starch which will produce pastes that will withstand several freeze-thaw cycles and prevent syneresis (weeping) occurs. Wide applications are in foods as texturing agent and provide good freeze-thaw stability. Extended applications in food industry are found by acetylated starch in conjunction with cross-linked starch.

8. Dextrin

Dextrin (E1400) is formed by roasting the starch with hydrochloric acid. Dextrination is the heating of powdered starch, mostly in the presence of small amounts of acids, at different temperatures and with different reaction times. Dextrin is used as adhesives in paper and textile based industry.

9. Grafted starch

Grafted starches are produced by free radical copolymerization with ethylenically unsaturated monomers. Starch grafted with synthetic polymers is most utilized tarches from different botanical origins were grafted with 1, 3 butadiene, styrene, acrylamide, acrylonitrile and Meth acrylic acid using free redox reaction.

10. Starch ethers

Starch ethers are produced by a nucleophilic substitution reaction with an ethylenically unsaturated monomer, followed by acid-catalyzed hydrolysis for viscosity adjustment.

11. Physically modified starch

Native starch can be modified with mechanical treatment, using spray drying technique, annealing technique

12. Enzyme modified starch

Enzyme-treated starch which includes maltodextrin, cyclodextrin Starch modified with amylase enzyme produces derivative with good adhesion property and mainly used in coating the food with colorant.

References

Lynch, M.J.(1992).Panning: An Overview. Manufacturing Confectioner, 72(5):59-64.

Rabinovitch, K and Benedict, S.2007. Method of Chocolate Coating soft confectionery centers, US Patent 7232584B2.

Talbot, G. (2009). Science and technology of enrobed and filled chocolate, confectionery and bakery products. Woodhead Publishing Limited, Abington Hall, Granta Park, Great Abington, Cambridge CB216AH,UK.

10
Commercial Processing of Chocolate

Chocolate is a key ingredient in many foods such as milk shakes, candy bars, cookies and cereals. It is ranked as one of the most favourite flavours in North America and Europe. Despite its popularity, most people do not know the unique origins of this popular treat. Chocolate is a product that requires complex procedures to produce. The process involves harvesting coca, refining coca to cocoa beans, and shipping the cocoa beans to the manufacturing factory for cleaning, conching and grinding. These cocoa beans will then be imported or exported to other countries and be transformed into different type of chocolate products.

In history, 1828 marked the "modern era" of chocolate making when Dutch chocolate maker Conrad J. van Houten patented an inexpensive method for pressing the fat from roasted cacao beans along with other processes to create a fine powder known as "cocoa". The powder was then treated with alkaline salt that helped the powder mix with water easily. The creation of powdered chocolate made it easier to mix with water, sugar and a possible combination of other ingredients to make chocolate in solid form. Many other chocolate makers began to build on Van Houten's success to make a variety of chocolate products. In 1894, English chocolate maker Joseph Storrs Fry produced what was arguably the world's first eating chocolate"

Harvest and Fermentation

Cocoa beans are produced from Theobroma cocoa trees. Cocoa trees are evergreens that do best within 20 degrees of the equator, at altitudes of between 100 (30.48 centimeters) and 1,000 (304.8 centimeters) feet above sea level. Native to South and Central America, the trees are currently grown on commercial plantations in such places as Malaysia, Brazil, Ecuador, and West Africa. West Africa currently produces nearly three quarters of the world's 75,000 ton annual cocoa bean crop, while Brazil is the largest producer in the Western Hemisphere. Because they are relatively delicate, the trees can be harmed by full sun, fungi, and insect pests. To minimize such damage, they are usually planted with other trees such as rubber or banana. The other crops afford protection from the sun and provide plantation owners with an alternative income if the cocoa trees fail.

Cacao trees produce buds on a continuous basis—this can be year-round in subtropical areas, such as Central America or it can be tied to the rain cycle, as it is in Africa. Fluctuations in growth cycle and harvest can occur because of changes in climate conditions. The pods, the fruit of the cocoa tree, are 6-10 inches (15.24–25.4 centimeters) long and 3–4 inches (7.62–10.16 centimeters) in diameter. Most trees bear only about 30–40 pods, each of which contains between 20 and 40 inch-long (2.54 centimeters) beans in a gummy liquid. The pods ripen in three to four months, and, because of the even climate in which the trees grow, they ripen continually throughout the year. However, the greatest number of pods are harvested between May and December.

Of the 30–40 pods on a typical cacao tree, no more than half will be mature at any given time. Only the mature fruits can be harvested, as only they will produce top quality ingredients. After being cut from the trees with machetes or knives mounted on poles (the trees are too delicate to be climbed), mature pods are opened on the plantation with a large knife or machete. The beans inside are then manually removed.

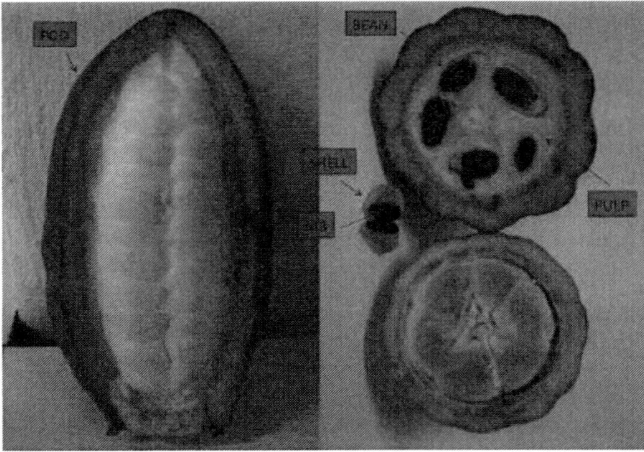

(a) On a Theobroma Cacao [2] (b) An opened cacao pod with pulp, beans and nib labeled [1]

Fig 10.1: Cacao pod, bean and nib

Still entwined with pulp from the pods, the seeds are piled on the ground, where they are allowed to heat beneath the sun for several days (some plantations also dry the beans mechanically, if necessary). Enzymes from the pulp combine with wild, airborne yeasts to cause a small amount of fermentation that will make the final product even more appetizing. During the fermenting process, the beans reach a temperature of about 125 degrees Fahrenheit (51 degrees Celsius). This kills the embryos, preventing the beans from sprouting while in transit; it also stimulates decomposition of the beans' cell walls. Once the beans have sufficiently fermented, they will be stripped of the remaining pulp and dried. Next, they are graded and bagged in sacks weighing from 130–200 pounds (59.02–90.8 kilograms). They will then be stored until they are inspected, after which they will be shipped to an auction to be sold to chocolate makers.

The pods are carefully broken open to release the cacao beans, which are embedded in a moist, fibrous, white pulp. The beans and pulp are scooped out quickly and either heaped in a pile on mats or banana leaves and covered, or put into a bin or box with a lid.

(a) Fermenting in a box

(b) Drying in the sum after fermentation

Fig 10.2: Fermenting boxes and Drying yard for beans

Fermentation occurs when the pulp surrounding the cacao bean is converted into alcohol by the yeasts present in the air and the heat generated by the pile or box. The beans are mixed gently during this process to introduce oxygen into the pile or box, which turns the alcohol into lactic and acetic acid. Slits or holes in the box allow the resulting liquid with its alcohol content to slowly leak out of the pile of beans during the fermentation process, leaving just the beans.

Germination in the cacao bean is killed by the high temperatures produced during the fermentation process. The beans gather moisture from the environment and plump. Their flavor begins to change from mainly bitter to the beginnings of the complex flavor called chocolate. The fermentation process can take up to eight days, depending on the species of cacao bean. Better fermentation results in better flavor and requires less roasting time to bring out that flavor.

Drying and Storage

The cocoa beans, as they are now called after fermentation, come out of this process with high moisture content. In order to be shipped or stored, they must be dried. The drying process differs, depending on the climate or size of the plantation. Cocoa beans can be dried out in the sun on trays or mats where the climate permits. Sun-drying usually happens in smaller plantations in drier environments. In tropical areas, where daily rainfall is the norm, the beans can be dried in sheds, as long as there is enough air circulating around the flats of beans. The use of wood fire to speed the drying process is disliked by bulk chocolate manufacturers and chocolatiers, as the process leaves the beans with a smoky taste.

Once the moisture percentage in the cocoa beans has reached 6 to 7 percent, they are sorted and bagged. The sorting process is very important because the cocoa beans are classified and sold in the industry by their size and quality. The bagged cocoa beans are then loaded on ships to be delivered to chocolate manufacturers.

Each chocolate manufacturer has a closely guarded "secret recipe" for each chocolate product that it produces. This secret begins with the type and quality of the cocoa beans used.

Testing, Cleaning, and Roasting

When the selected cocoa beans arrive at the manufacturing plant they go through a very extensive sampling and testing procedure. Sample cocoa beans are tested for size and defects, such as insects or mold, and then converted into chocolate liquor, which is evaluated for flavor and aroma by company tasters. Once the testing is complete and the shipment is accepted by the manufacturer, the beans are thoroughly cleaned to remove any foreign matter. The cocoa beans then go into the roaster for anywhere from 10 to 35 minutes.

Cracking (or Fanning) and Grinding

While roasting, the shell of the cocoa bean separates from the bean kernel and is removed in the first step of the cracking or fanning process .The dried beans are cracked and a stream of air separates the shell from the nib, the small pieces used to make chocolate. The rotating blades crack the beans into pieces of nib and shell. The beans are cracked (not crushed) by being passed through serrated cones. The cracked beans are now called *cocoa nibs.*

As the shell is dry and lightweight, it can be winnowed from the cocoa nib. Winnowing is done by exposure to a current of air, so that the shells are blown free of the heavier nibs. The nibs contain approximately 53 percent cocoa butter, depending on the cacao species.

Fig 10.3 i) Cracked beans with shells and ii) Cleaned nibs

Roasting

The nibs are roasted in special ovens at temperatures between 105-120 degrees Celsius. The actual roasting time depends on whether the end use is for cocoa or chocolate. During roasting, the cocoa nibs darken to a rich, brown colour and acquire their characteristic chocolate flavour and aroma. This flavour however, actually starts to develop during fermentation. Roasting also causes the shells to open and break away from the nibs (the meat of the bean). This separation process can be completed by blowing air across the beans as they go through a giant winnowing machine called a cracker and fanner, which loosens the hulls from the beans without crushing them. The hulls, now separated from the nibs, are usually sold as either mulch or fertilizer. They are also sometimes used as a commercial boiler fuel.

Next, the roasted nibs undergo *broyage,* a process of crushing that takes place in a grinder made of revolving granite blocks. The design of the grinder may vary, but most resemble old-fashioned flour mills. The final product of this grinding process, made up of small particles of the nib suspended in oil, is a thick syrup known as chocolate liquor.

Grinding or Refining

The first grind of the beans is usually done in a milling or grinding machine such as a melangeur. The nibs are ground or crushed to liquefy the cocoa butter and produce what is now called *chocolate liquor* or *chocolate liquid.*

For the second refining process, most chocolate manufacturers use a roll refiner or ball mill, which has two functions: to further reduce the particle size of the cocoa mass (and any other ingredients, such as sugar or milk powder) and to distribute the cocoa butter evenly throughout the mass, coating all the particles.

The rolling process itself creates heat that melts and distributes the cocoa butter. As well as the flavor of the chocolate, manufacturers must decide on the particle size for each of their chocolates. This is the first step to developing chocolate's smooth and creamy mouth-feel.

Different percentages of cocoa butter are removed or added to the chocolate liquor. Cocoa butter carries the flavor of the chocolate and produces a cooling effect on your tongue that you might notice when eating dark chocolate. Also, depending on the chocolate flavor desired, some or all of the following ingredients are added: sugar, lecithin, milk or cream powder or milk crumb (used to produce a caramel-like taste in milk chocolate), and spices such as vanilla. The formula the chocolate manufacturer develops for combining specific ingredients with the chocolate liquor is what gives the chocolate its unique taste.

The cocoa mass is pressed in powerful machines to extract the cocoa butter, vital to making chocolate. The solid blocks of compressed cocoa remaining after extraction (press cake) are pulverized into a fine powder to produce a high-grade cocoa powder for use as a beverage or in cooking. The cocoa mass, cocoa butter and cocoa powder are then quality inspected

Conching

This process develops the flavor of the chocolate liquor, releasing some of the inherent bitterness and gives the resulting chocolate its smooth, melt-in-your-mouth quality. The conch machine has rollers or paddles that continuously knead the chocolate liquor and its ingredients over a period of hours or days depending on the flavor and texture desired by the manufacturer. Thick chocolate is needed for moulded blocks, while a thinner consistency is used for assortments and covering bars.

Both milk and dark chocolate undergo the same final special production stages - refining, conching and tempering - which produce the famous smoothness, gloss and snap of chocolate. Conching involves mixing and beating the semi-liquid mixture to develop the flavour, removing unwanted volatile flavours and reducing the viscosity and particle size.

Tempering and Forming Chocolate

For the last two steps in the chocolate process, the conched chocolate mass is tempered and molded into bulk bars or it may go into another production cycle to produce specialized retail products, such as coated-candy centers and molded items.

On the most basic level, tempering is necessary because the particles that make up a chocolate bar can arrange themselves in many different ways. The different arrangements of the chocolate particles on a molecular level create different physical properties of the final chocolate on a much larger scale. Chocolate with the correct molecular arrangement (referred to as Form V chocolate) is dark brown, glossy, and makes a satisfying snap when broken .Chocolate with an incorrect molecular arrangement (Form IV chocolate) is lighter in color, matte, and will crumble when broken instead of snapping]. Mistempered chocolate will also exhibit an unsightly white coating called fat bloom .

(a) Properly tempered chocolate is dark brown and

(b) Poorly tempered chocolate is lighter and crumbles

(c) Poorly tempered chocolate also exhibits ugly white impurities known as fat bloom

Fig 10.4 Various forms of Tempered chocolate

Tempering is the final crucial and complex stage which involves mixing and cooling the liquid chocolate under carefully controlled conditions to ensure that the fat in the chocolate crystallizes in its most stable form. Highly sophisticated machinery has been developed for this process, which is one of the skills of the chocolatier. Tempered chocolate is used in a number of ways to produce our famous brands. Blocks of solid chocolate, including bars with added ingredients such as nuts and raisins, are known in the industry as 'moulded' products. Tempered chocolate is poured into bar-shaped moulds, shaken and cooled, then the moulded blocks continue to high speed wrapping plants.

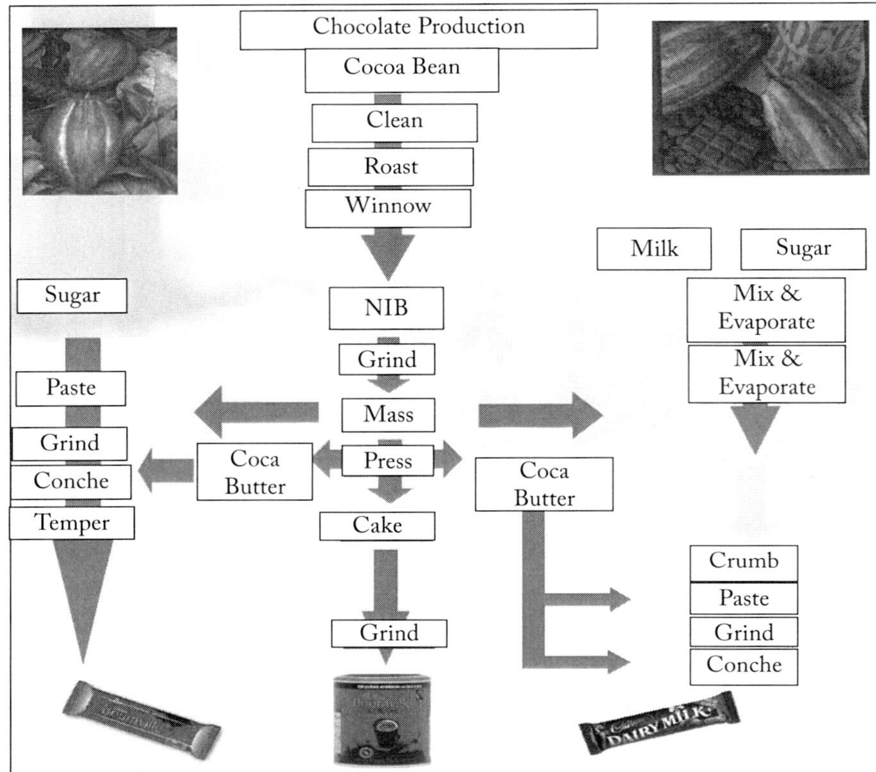

Fig 10.5 Overall processing of bean to chocolate

Making Chocolate Candy

If the chocolate being produced is to become candy, the press cake is remixed with some of the removed cocoa butter. The restored cocoa butter is necessary for texture and consistency, and different types of chocolate require different amounts of cocoa butter. The mixture now undergoes a process known as *conching,* in which it is continuously turned and ground in a huge open vat. The process's name derives from older vats, which resembled large conch shells. The conching process can last from between three hours to three days (more time is not necessarily better, however). This is the most important step in making chocolate. The speed and temperature of the mixing are critical in determining the quality of the final product.

Making Cocoa Powder

If the chocolate being produced is to be cocoa powder, from which hot chocolate and baking mixes are made, the chocolate liquor may be *dutched,* a process so-named because it was invented by the Dutch chocolate maker Conrad van Houten. In the dutching process, the liquor is treated with an alkaline solution, usually potassium carbonate, that raises its pH from 5.5 to 7 or 8. This darkens the color of the cocoa, renders its flavor more mild, and reduces the tendency of the nib particles to form clumps in the liquor. The powder that eventually ensues is called Dutch cocoa.

The next step in making cocoa powder is *defatting* the chocolate liquor, or removing large amounts of butter from it. This is done by further compressing the liquor between rollers, until about half of the fat from its cocoa beans has been released. The resulting solid material, commonly called *press cake,* is then broken, chopped, or crushed before being sifted to produce cocoa powder. When additives such as sugar or other sweeteners have been blended, this cocoa powder becomes a modern version of chocolate.

References

Afoakwa, Emmanuel Ohene. Chocolate Science and Technology. Hoboken: Wiley-Blackwell. 2010.

Head, Brandon. The Food of the Gods. http://www.gutenberg.org/files/16035/16035-h/16035-h.htm

Types of Chocolate. Wikipedia. http://en.wikipedia.org/wiki/Types_of_chocolate

11

Types of Confectionery Hard Boiled Sweets, Aerated Confectionery, Granulated, Sugar Panning Tablets, Cream Pastes and Lozenges

Types of Confectionery

Confectionery also called sweets or candy is sweet food product. Confectionery is divided into two broad and somewhat overlapping categories, Bakers' Confections and Sugar Confections. Confections are low in micronutrients and protein but high in calories. They may be fat-free foods, although some confections, especially fried doughs, are high-fat foods. Many confections are considered to give empty calories.

Baker's Confectionery

Baker's confectionery also called flour confections includes principally sweet pastries, cakes, and similar baked goods. In the Middle East and Asia, flour-based confections are more dominant. Baker's confectionery includes sweet baked goods, especially those that are served for the dessert course. Baker's confections are sweet foods that feature flour as a main ingredient and are baked. Major categories include cakes, sweet pastries, doughnuts, scones, and cookie.

Fig 11.1 Picture showing varieties of baker's confectioneries

Sugar Confectioneries

Sugar confectionery includes sweets, candied nuts, chocolates, chewing gum, sweetmeats, pastillage, and other confections that are made primarily of sugar. In some cases, chocolate confections (confections made of chocolate) are treated as a separate category, as are sugar-free versions of sugar confections. The words candy (US and Canada), sweets (UK and Ireland), and lollies (Australia and New Zealand) are common words for the most common varieties of sugar confectionery. Sugar confections include sweet, sugar-based foods, which are usually eaten as snack food. This includes sugar candies, chocolates, candied fruits and nuts, chewing gum, and sometimes ice cream. Confections are defined by the presence of sweeteners. These are usually sugars, but it is possible to buy sugar-free sweets, such as sugar-free peppermints. The most common sweetener for home cooking is table sugar, which is chemically a disaccharide called sucrose. Hydrolysis of sucrose gives a mixture called invert sugar, which is sweeter and is also a common commercial ingredient. Commercial confectionery is sweetened by a variety of syrups obtained by hydrolysis of starch. These sweeteners include all types of corn syrup Specially formulated chocolate has been manufactured in the past for military use as a high density food energy source.

Fig 11.2 Picture showing various sugar confectioneries

Types of Bakers Confectioneries

Cakes: is a form of sweet dessert that is typically baked. Typical cake ingredients are flour, sugar, eggs, butter or oil, a liquid, and leavening agents, such as baking soda and/or baking powder. Common additional ingredients and flavourings include dried, candied or fresh fruit, nuts, cocoa, and extracts such as vanilla, with numerous substitutions for the primary ingredients. Cakes can also be filled with fruit preserves or dessert sauces (like pastry cream), iced with butter cream or other icings, and decorated with marzipan, piped borders, or candied fruit.

Sweet Pastries: Pastry is dough of flour and water and shortening that may be savoury or sweetened. Sweetened pastries are often described as bakers' confectionery. The word "Pastries" suggests many kinds of baked products made from ingredients such as flour, sugar, milk, butter, shortening, baking powder, and eggs

Doughnuts: is a type of fried dough confectionery or dessert food. Doughnuts are usually deep-fried from a flour dough, and typically either ring-shaped or without a hole, and often filled. Other types of batters can also be used, and various toppings and flavorings are used for different types, such as sugar, chocolate, or maple glazing.

Scones: scone is a single-serving cake or quick bread. They are usually made of wheat, barley or oatmeal, with baking powder as a leavening agent, and are baked on sheet pans. They are often lightly sweetened and are occasionally glazed with egg wash. The scone is a basic component of the cream tea or Devonshire tea. It differs from a teacake and other sweet buns, which are made with yeast.

Cookies: Are small, flat, sweet, baked good, usually containing flour, eggs, sugar, and either butter, cooking oil or another oil or fat. It may include other ingredients such as raisins, oats, chocolate chips or nuts.

Cakes Dough Pastry Doughnuts

Scones Cookies

Types of Sugar Confectioneries

Sugar confectionery items include sweets, lollipops, candy bars, chocolate, cotton candy, and other sweet items of snack food. Some of the categories and types of sugar confectionery include the following:

Caramels: Derived from a mixture of sucrose, glucose syrup, and milk products. The mixture does not crystallize, thus remains tacky.

Chocolates: Bite-sized confectioneries generally made with chocolate.

Divinity: A nougat-like confectionery based on egg whites with chopped nuts.

Dodol: A toffee-like food delicacy popular in Indonesia, Malaysia, and the Philippines

Dragée: Sugar-coated almonds and other types of sugar panned candy.

Fondant: Prepared from a warm mixture of glucose syrup and sucrose, this is partially crystallized. The fineness of the crystallites results in a creamy texture.

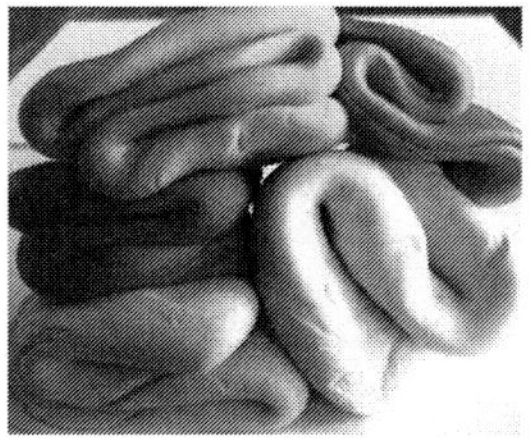

Fudge: Made by boiling milk and sugar to the soft-ball stage. In the US, it tends to be chocolate-flavored.

Halvah: Confectionery based on tahini, a paste made from ground sesame seeds.

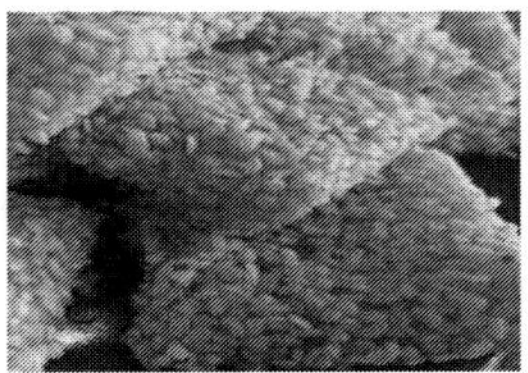

Hard candy: Based on sugars cooked to the hard-crack stage. Examples include suckers (known as boiled sweets in British English), lollipops, jawbreakers (or gobstoppers), lemon drops, peppermint drops and disks, candy canes, rock candy, etc. Also included are types often mixed with nuts such as brittle. Others contain flavorings including coffee such as Kopiko.

Ice cream: Frozen, flavoured cream, often containing small pieces of chocolate, fruits and/or nuts.

Jelly candies: Including those based on sugar and starch, pectin, gum, or gelatin such as Turkish delight (lokum), jelly beans, gumdrops, jujubes, gummies, etc.

Liquorice: Containing extract of the liquorice root. Chewier and more resilient than gum/gelatin candies, but still designed for swallowing. For example, Liquorice all sorts has a similar taste to star anise.

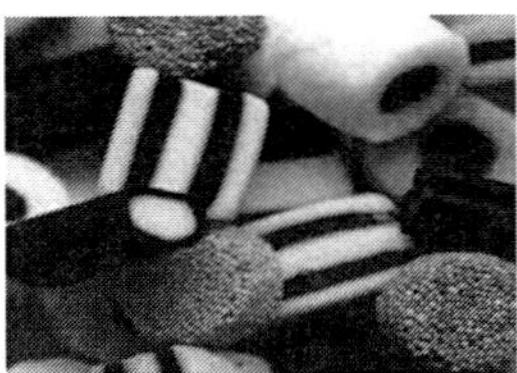

Marshmallow: "Peeps" (a trade name), circus peanuts, fluffy puff, Jet-Puffed Marshmallows

Marzipan: An almond-based confection, doughy in consistency, served in several different ways.

Mithai: A generic term for confectionery in India, typically made from dairy products and/or some form of flour. Sugar or molasses are used as sweeteners.

Tablet: A crumbly milk-based soft and hard candy, based on sugars cooked to the soft ball stage. They come in several forms, such as wafers and heart shapes. Not to be confused with tableting, a method of candy production.

Toffy or chews: A candy that is folded many times above 120 °F (50 °C), incorporating air bubbles thus reducing its density and making it opaque. Chocolate means the confectionery product characterized by the presence of cocoa bean derivatives: - (a) Prepared from a minimum of 200 g/kg of cocoa bean derivatives; and (b) Which contains no more than 50 g/kg of edible oils, other than cocoa butter or dairyfats. (Source: Australia New Zealand Food Standards Code - Standard 1.1.2 - Supplementary Definitions for Foods) Principles of Sugar Confectionery Production Sugar confectionery refers to a large range of food items, commonly known as sweets.

Boiled sweets, toffees, marshmallows, and fondant are all examples. Sweets are a non-essential commodity, but are consumed by people from most income groups. The variety of products is enormous, ranging from cheap, individually-wrapped sweets, to those presented in boxes with sophisticated packaging.

By varying the ingredients used, the temperature of boiling, and the method of shaping, it is possible to make a wide variety of products. In all cases, however, the principle of production remains the same and is outlined below:

- balance the recipe
- prepare the ingredients
- mix together the ingredient
- boil the mixture until the desired temperature has been reached
- Cool
- Shape
- Pack.

Sweets containing high concentrations of sugar (sucrose) may crystallize either during manufacture or on storage (commonly referred to as graining). Although this may be desirable for certain products (such as fondant and fudge), in most other cases it is seen as a quality defect. When a sugar solution is heated, a certain percentage of sucrose breaks down to form 'invert sugar'. This invert sugar inhibits sucrose crystallization and increases the overall concentration of sugars in the mixture. This natural process of inversion, however, makes it difficult to accurately assess the degree of invert sugar that will be produced. Variations in boiling temperature can make a difference between a sticky, cloudy sweet or a dry, clear sweet.

An accurate way of measuring the temperature is to use a sugar thermometer. Other tests can be used to assess the temperature (for example, toffee temperatures can be estimated by removing a sample, cooling it in water, and examining it when cold). The temperatures are known by distinctive names such as 'soft ball', 'hard ball' etc., all of which refer to the consistency of the cold toffee.

Type of sweet	Temperature range for boiling (Degree C)
Fondants	116–121
Fudge	116
Caramels and regular toffee	118–132
Hard toffee (e.g., butterscotch)	146–154
Hard-boiled sweets	149–166

Moisture Content

The water left in the sweet will influence its storage behavior and determine whether the product will dry out, or pick up, moisture. For sweets which contain more than 4 per cent moisture, it is likely that sucrose will crystallize on storage. The surface of the sweet will absorb water, the sucrose solution will subsequently weaken, and crystallization will occur at the surface - later spreading throughout the sweet.

Added ingredients

The addition of certain ingredients can affect the temperature of boiling. For example, if liquid milk is used in the production of toffees, the moisture content of the mixture immediately increases, and will therefore require a longer boiling time in order to reach the desired moisture content. Added ingredients also have an effect on the shelf-life of the sweet. Toffees, caramels, and fudges, which contain milk-solids and fat, have a higher viscosity, which controls crystallization. On the other hand, the use of fats may make the sweet prone to rancidity, and consequently the shelf-life will be shortened.

Types of Sweets

Fondants and creams

Fondant is made by boiling a sugar solution with the optional addition of glucose syrup. The mixture is boiled to a temperature in the range of 116–121 °C, cooled, and then beaten in order to control the crystallization process and reduce the size of the crystals. Creams are fondants which have been diluted with a weak sugar solution or water. These products are not very stable due to their high water content, and therefore have a shorter shelf-life than many other sugar confectionery products. Both fondants and creams are commonly used as soft centres for chocolates and other sweets.

Gelatin sweets

These sweets include gums, jellies, pastilles, and marshmallows. They are distinct from other sweets as they have a rather spongy texture which is set by gelatin.

Toffee and caramels

These are made from sugar solutions with the addition of ingredients such as milk solids and fats. Toffees have lower moisture content than caramels and consequently have a harder texture. As the product does not need to be clear, it is possible to use unrefined sugar such as jaggery or gur, instead of white granular sugar.

Hard-boiled sweets

These are made from a concentrated solution of sugar which has been heated and then cooled to form a solid mass containing less than 2 per cent moisture. Within this group of products there is a wide scope to create many different colours, flavours and shapes through the use of added flavourings and colourings.

Hard Boiled Candy Manufacturing Process

Hard boiled confectionery(HBC) available in the different form such as Candy, Hard boiled sweets, Lozenges, Sugar Drops due to its popularity among all age groups, the different market player such as **Lotte, Mars, Krafts (Mondelez), Surya Foods (Priyagold)** are important manufacturers of HBC in India. In 2015, **Dharmpal Satypal Group, Commonly known as D.S. Group** who launched **Pulse candy** which hit the market with 100 Cr. in 8 Months of its launch.

Processing of Hard Boiled Candies

Confectioners start the processing of Hard Boiled candies with **sucrose, glucose** syrup (Confectioners syrup), with different fruit flavoring and coloring material and acids. Each ingredient plays a specific role like sucrose gives sweetness, texture and increase the shelf life of the product. Liquid glucose help to inhibit Crystallization of sucrose, contribute to the texture of the product. Flavorings Plays important role as imparts desirable flavor to HBC. Today people are shifting toward fusion flavors in which blend of 2 Flavors is used. The coloring is mostly to attract the consumer, also help to distinguish the product. Acids used to control sugar inversion to make invert sugar can have dual roles depending on the candy and pH level.

Process

Processing of HBC starts with ingredient mixing in large vacuum cooker or batch cooker. Batch cooker helps to Mix and heat the candy mass at desired temperature which is necessary for candy making. Different size and forms of batch cookers are available in the market. For continuous process Vacuum cooker with Feeding pump is attached to increase the efficiency of the system. Temperature control and end point of the process is very important. The temperature of boiling directly affects the final sugar concentration and moisture content of the confectionery product.

Once the HBC is up to temperature remove it from the heat and pour it straight onto the candy cooling table. While the batch is still liquid on the table we can add our colors and flavor. (Adding them here as opposed to in the pan prevents colors and flavors leaking into the next batch) Add the color and flavor to the center of the batch and gently spread it out using a scraper. As the candy cools it will begin to form a skin on the underside during contact with the cooling table. We need to ensure the cooling happens evenly throughout the batch. To achieve this batch is scraped from the table and folded in on it. Experienced candy makers can do this using their hands only. If you're new to candy making then you may wish to use a scraper and use gloves. After each fold, you should leave it to cool and settle for about 30 seconds before repeating.

At this point, your candy should be firm but pliable. Some people will now pull air into the batch either by hand over a hook or with an automatic pulling machine.Pulling off the HBC bulks it out in size and helps develop colors. HBC roper or size cutting machine is used to get different size and shape of the HBC.

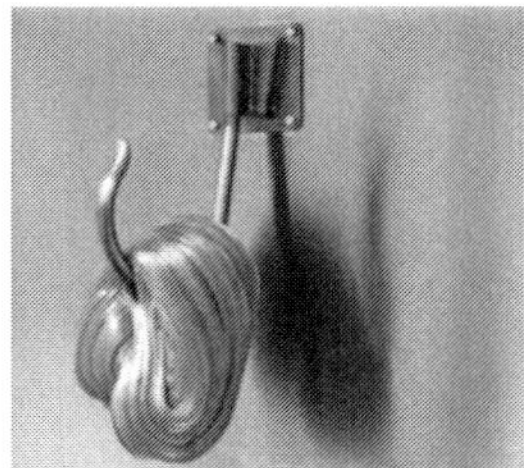

Fig 11.3 Candy Pulling Hook

11.3 Aerated confectionery

Aerated confectionery can be defined as an aerated gelled product containing a mixture of carbohydrates, mainly sugar and different types of glucose syrup, whipping and/or stabilizing agents, flavor and color.The aeration technique enables a liquid to be transformed into foam by incorporating a certain volume of air in the form of finely divided bubbles.

This technique causes:

- An increase in volume, together with a decrease in density
- A modification of the viscosity and fluidity of the aerated mass, leading to a better stability
- A modification of the texture and organoleptic characteristics of the finished products

Aeration of the product leads to:

- A shorter texture
- A modification in the mouthfeel
- A reduction of stickiness and cold flow
- A decrease in sweetness

Marshmallows are one of the earliest confections known to humankind. Today's marshmallows come in many forms, from solid (soft pillows dropped in cocoa or roasted on a stick) to semi-liquid (covered in chocolate or formed into chicks for Easter) to the creme-like (used as a base in other candies or as an ice cream topping). In essence, all marshmallows are aerated candies.

History

Originally, however, marshmallows were made from the root sap of the marsh mallow *(Althaea officinalis)* plant. It is a genus of herb that is native to parts of Europe, north Africa, and Asia. Marsh mallows grow in marshes and other damp areas. The plant has a fleshy stem, leaves, and pale, five-petaled flowers. The first marshmallows were made by boiling pieces of the marsh mallow root pulp with sugar until it thickened. After it had thickened, the mixture was strained and cooled. As far back as 2000 B.C. , Egyptians combined the marsh mallow root with honey. The candy was reserved for gods and royalty.

Raw Materials

Marshmallows are made from only a few ingredients, which fall into two main categories: sweeteners and emulsifying agents. Sweeteners include corn syrup, sugar, and dextrose. Proportionally, there is more corn syrup than sugar because it increases solubility (the ability to dissolve) and retards crystallization. Corn starch, modified food starch, water, gum, gelatin, and/or whipped egg whites are used in various combinations. The resulting combination gives the marshmallows their texture. They act as emulsifying agents by maintaining fat distribution and providing the aeration that makes marshmallows puffy. Gum, obtained from plants, also can act as an emulsifier in marshmallows, but it is also important as a gelling agent.

Most marshmallows also contain natural and/or artificial flavoring. If they are colored marshmallows, the color usually comes from an artificial coloring.

The Manufacturing Process

Cooking

- A solution is formed by dissolving sugar and corn syrup in water and boiling it. Egg whites and/or gelatin is mixed with the sugar solution. Then the ingredients are heated in a cook kettle to about 240 °F (115 °C). The resulting mixture is passed through a strainer to remove extraneous matter.
- In the pump, the mixture is then beaten into foam to two or three times its original volume. At this stage, flavoring can be added.

Forming

- The heated mixture is transferred to a heat exchanger. Air is pumped into the mixture. The mixture cools in a tempering kettle, passes through another filter, and continues on to the "hill." Marshmallows are extruded through a machine or deposited onto bands.

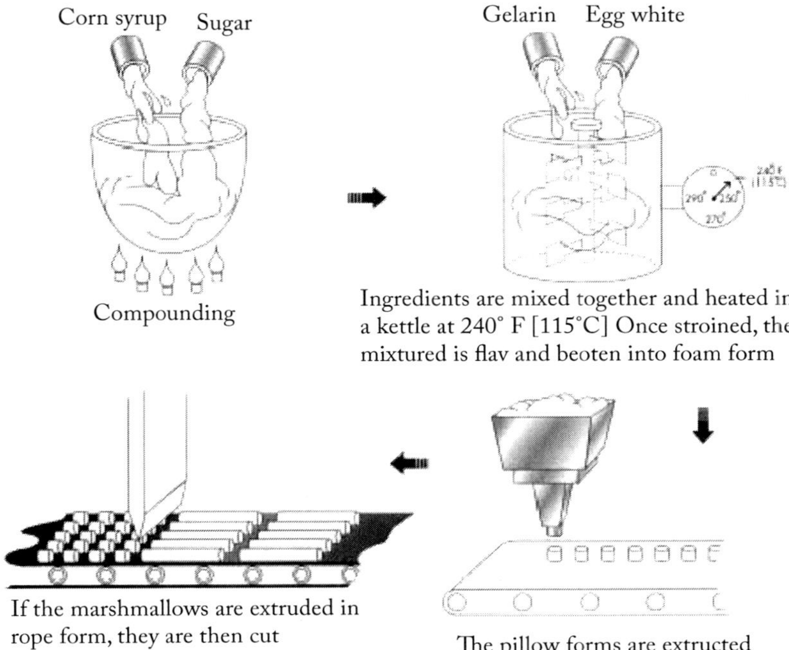

Fig 11.4 Marshmallow processing

The extrusion process involves the foam being squeezed through a die to produce marshmallow's familiar pillow shape. Usually, they get a coating of corn starch to counter stickiness and help maintain their form after they have been extruded. Sometimes the pillows are formed into a rope of pillows. If so, they are cut and dried on a rubber conveyor belt.

Cooling

After the pillows are formed, they are sent through a cooling drum, where excess starch is removed. They also are cooled enough to be packaged.

Packaging

After the pillows have cooled, they are weighed and packaged. Before being put in cases, some manufacturers pass their product through a metal detector. The case is code dated and shipped to retail stores.

Lozenge Mufacturing Process

The term "lozenge" is somewhat confusing because it does not refer to a specific type of candy. Products may be labeled "lozenges" that are made by the cut process, by compression or of hard candy or hard jelly gums.The most common connotation of this name is that lozenges usually have a round shape and some benefit other than just taste – breath freshening, relief of a sore throat or clearing of the sinuses.

Cut lozenges are made from a dough that is basically powdered sugar mixed with an adhesive syrup, colored and flavored. A typical formulation would be: Powdered sugar (10X) -200 lbs, Gum arabic (acacia) - 8 lbs, Water - 9 lbs, Gum tragacanth - 8 oz, Water - 3 lbs, Gelatin (150 bloom) - 12 oz, Hot water - 3 lbs, Corn syrup (regular 42/43) -17 lbs, Corn starch - 8 lbs and Color and flavor as desired

The gum arabic solution is prepared in advance and is used at room temperature. If gum arabic is the only adhesive used, the lozenges will be rather hard and brittle. The gum tragacanth requires several hours of soaking to absorb its portion of water. This adhesive is tenacious and provides better strength than gum arabic alone. Gelatin is stirred into its portion of hot water, just prior to use, to aid adhesion and provide flexibility. The corn syrup is preferably warmed to aid dispersion. It contributes to adhesion, but also retains some moisture so that the lozenges are not hard and brittle.

Depending on the fineness of the sugar, more or less of the gum arabic solution may be needed. Very fine sugar will have a larger surface area of particles to be wetted, and this will require more fluid. Conversely, a coarser grade of sugar will need less adhesion solution. The required texture is a soft, plastic paste that can be rolled into a thin sheet. The double sigma-blade dough mixer is generally used. Normally, this need not be jacketed as mixing is at room temperature. Other efficient, heavy duty mixers may be used. The usual sequence is to place all of the sugar and starch into the mixer and then add the adhesives, color and flavor while mixing until a homogeneous mass is produced. The next step is to form the shaped lozenges.

The equipment that is normally used forms a continuous sheet of about 1/2-inch thickness by means of a horizontal multiple-screw extruder. The sheet is then rolled down to the required thickness by passing through multiple pairs of horizontal rollers. The sheet is heavily dusted with dry starch to prevent sticking to the rollers. Next the sheet passes beneath a set of vertical cutters that stamp out the required shape as well as imprint a product name or company logo on the surface. The shaped lozenges are delivered onto starch-dusted trays and the web is reworked through the extruder. In the preceding formula the moisture content is approximately 7.5 percent. The lozenges must be dried to between one and two percent moisture to make them hard. The lozenges are spread in trays that have perforated bases to assist the removal of moisture. There should be a sufficient gap between the trays to permit free circulation of drying air. The trays are held in rooms where the temperature and relative humidity are controlled to maximize the rate of drying.

Depending on formulation and thickness, drying may take days. Drying rooms should have fans to circulate air between the trays. If drying is too rapid, a hard skin may form that will delay drying of the inside. If the temperature is too high, the lozenges may expand and lose their flat appearance. At a high temperature, there may be much loss of volatile flavoring. Typical drying room conditions will have a temperature of about 100-110 degrees F and a differential of about 25 percent between the equilibrium relative humidity of the product and relative humidity of the air. There must be good air circulation, but not too much to blow starch from the trays.

Sugar Panning Tablets

Panning is an art of framing layers on candy-based centers in a very controlled way . Sugar panning means application of a thin coat of sucrose solution to every individual tumbling centers followed by evaporation of moisture so that a thin layer of crystallized sugar is formed Chocolate panning is a generic term for the continuous application of a fat based coating to a prepared center

Panning process is the controlled build-up of a center through application of successive layers either of solid or liquid coating material in a revolving pan, with or without the use of warm or cool air to dry or set the coating Panning process consists of the following three steps: a. Pretreatment of the centre b. Chocolate/compound panning c. Polishing and sealing

Sugar panning is a process of building up a layer by layer coating of sugar on centers The coating may be hard or soft, depending on the thickness, sugar composition and method of manufacture and the resultant sweets are called dragees.

In hard panning, the centres are tumbled in the pan and a sugar syrup is applied Both together, the rotation of the pan and the tumbling of the centres results into spreading of the syrup over the surface of the centres into a thin layer The evaporation of water in the sugar syrup causes crystallization of sugar and increasing the temperature reduces the rate of crystallization. The layers applied are only 10-14 μm thick, and as they are so thin, contours of the product occurs. The centres have to be coated with a concentrated sugar syrup for hard panning. Some centres, such as nonpareils, quickly take the sugar syrup coating, while others such as nuts or chewing gum have a hydrophobic surface, need some pretreatment The syrup is dosed in it is called 'wetting' and 'engrossing' when the coating is built up Hard panned confections have a hard crystalline coating

Soft panning syrup is not intended to crystallize The syrup used can either be an all glucose syrup or a 50:50 mixture of sucrose and glucose syrup Then caster or milled sugar is added which dissolves in the water of the syrup rather than evaporating the water as in hard panning. Any excess of sugar convert the syrup from a non-crystallizing syrup to a crystallizing one. Then the centres are removed from the pan and placed on trays for drying. Soft panning is a cold process and it does not use drying air A product that has been soft panned can be finished by dusting with milled sugar followed by a number of hard panned coats e.g. jelly beans and dolly mixture components. Soft panning applies a thick, soft layer to centers such as moulded jelly beans or chews.

Sugar panned confectionery products include

Soft panned: jelly beans/eggs (gourmet, seasonal, sports beans), gel center products (fruit sours, candy fruit rocks), sugar coated marshmallow eggs, some imperials

Hard panned: chewing gum, licorice, nuts, compressed tablets or high solids chewy candy, non-pareils

Granulation for tablet making

Tablets are formed by compressing powder in a dye. To form a tablet successfully, the powder must be free flowing and yet to be capable of bonding under pressure. Certain powders already posses these characteristics and are termed "directly compressible" materials. Other powders may be formed into tablets, but must first undergo a granulation stage.

The aim of granulation process is to produce a free flowing material, suitable for compression. There are several ways of achieving:

1. **Wet Granulation**: Wet granulation is suitable for most materials but expensive in terms of manufacturing space, time and energy. The powder to be granulated is screened to a uniform particle size. The granulating solution is a binder dissolved in water which will glue particle together to form granules. Binders are materials such as gum Arabic, gelatin, starches and alginates. The correct level of binder addition is such that the material may be compressed in the hand to form a ball which will not crumble or sticky when broken apart. The mixing may take up to one hour. The granules are dried on trays up to 24 hours at 50–60c.
2. **Fluidized Bed Granulation**: With this method the bed of powder is fluidized in an air stream and sprayed with binder solution. The powder agglomerates into granules which are subsequently dried in the air stream. Once the granules are dry lubricant and flavour may be added and mixed by a further period of fluidization.
3. **Sluggng**: It is also termed as "double compression". This is particularly suitable for moisture sensitive materials such as effervescent tablets. The powder is fed into a large die in a heavy duty tablet press. The material is compressed using flat faced punches into a rough tablet. This is done

slowly to allow air to escape. Since a binder is not added to hold the granules together the resultant tablets tend to be softer than those which have been wet granulated.

Cream Paste

Cream paste belongs to the family of sugar pastes. Cream paste is prepared by mixing icing sugar with mucilage of glucose syrup, gelatin, fat.

The residual moisture contents of cream paste are used as a means of quality determination:

(i) 5.5% moisture content, the product will be fresh, soft and short eating

(ii) 3.5% moisture content, the product content becomes stale, dry and firm.

Thus is very important that the cream paste make up is consistent and the moisture content constant. A cream paste consists of slid and liquid phases, in which the liquid phase is saturated with respect of sucrose. The equilibrium relative humidity of cream paste varies between 65 and 72, depending upon the recipe make up, under normal climatic conditions.

Cream paste ingredients

The main ingredients used in the manufacture of cream paste are: Sucrose, glucose syrup, invert sugar syrup, hard fat, gelatin, corn flour, coconut flour. These are various ingredients which are used to improve the flavour and modifies texture.

Manufacture

Cream paste is manufactured by mixing together icing sugar and a mucilage in a heavy-duty-Z-arm mixer, in which a homogenous mix is obtained. There are two main processes by which a cream paste is made:

1. With boiled mucilage
2. Non boiled mucilage

The method adopted depends on the equipment available and the type of texture required. The most used method for production of cream paste is continuous process of non boiled system. The non boiled method would use the following recipe:

Icing sugar	100
Glucose syrup	30
Invert sugar syrup	7
Hard palm kernel oil	7
Gelatin	1
Hot water	3

The glucose syrup, invert sugar syrup and the fat are heated together to 200 °F, and held until the fat gets dissolved. The gelatin solution is then added and mixed in, forming the mucilage. The mucilage is then transferred to a Z-arm mixer and the appropriate colours and flavours added, the icing sugar is fed in continuously as the Z arm revolve to give a uniform mix of a cream paste. The moisture content of the paste can be controlled by varying the amount of water used in the preparation of the gelatin solution.

The boiled method would use the following recipe:

Icing sugar	100
Granulated sugar	15
Glucose syrup	25
Invert sugar syrup	7
Hard palm kernel oil	8
Gelatin	1
Hot water	3

The granulated sugar, glucose syrup, and invert sugar syrup are heated in a sweet pan and boiled to 230°F. When the boiling has subsided, the fat and the gelatin solution are mixed in; the mucilage is transferred to a Z-mixer and the icing sugar added continuously, again to give a uniform mix of cram paste. The moisture content of the paste is controlled by the boiling temperature of the syrup during the preparation of the mucilage. Mostly cream paste is never left alone, but is layered with others colours of cream paste, with pectin jelly or with liquorice paste.

References

http://www.fao.org/WAIRdocs/x5434e/x5434e0a.htm
http://www.ldoceonline.com/dictionary/confectionery
https://www.britannica.com/topic/candy-food
International Food Information Service, ed. (2009). Dictionary of Food Science and Technology (2nd Ed.). Chichester, U.K.: Wiley–Blackwell. p.39
Rhea Lanting, M.S., Candy making Manual, University of Idaho, Extension Educator

12

Crystallized Confectionery and Chewing Gums

Chemically, sugar candies are broadly divided into two groups: crystalline candies and amorphous or non crystallized candies. Crystalline candies are not as hard as crystals of the mineral variety, but derive their name and their texture from their microscopically organized sugar structure, formed through a process of crystallization, which makes them easy to bite or cut into. Fudge, creams, and fondant are examples of crystalline candies. Amorphous candies have a disorganized crystalline structure. They usually have higher sugar concentrations, and the texture may be chewy, hard, or brittle. Hard candies, such as lollipops, caramels, nut brittles and toffees are all examples of amorphous candies, even though some of them are as hard as rocks and resemble crystals in their overall appearance. Crystalline candies are chemically described as having two phases, because the tiny, solid sugar crystals are suspended in a thick liquid solution. These are also called grained candies, because they can have a grainy texture. Amorphous candies are have only one phase, which is either solid or liquid, and do not have a grainy texture, so they may be called ungrained.

Commercially, candies are often divided into three groups, according to the amount of sugar they contain:

- 100% sugar (or nearly so), such as hard candies or creams
- 95% sugar or more, with up to 5% other ingredients, such as marshmallows or nougats, and
- 75–95% sugar, with 5–25% other ingredients, such as fudge or caramels. Each of these three groups contains both crystalline (grained) and amorphous (ungrained) candies.

Preparation of Crystalline and Non Crystalline Candies

Candy making is an exact science and an art, its success largely dependent on the knowledge of the science of sugar crystallization and timing. The goal in preparing soft, creamy and smooth textured crystalline candy is to develop numerous very fine nuclei in the sugar syrup solution. They are formed by: 1) controlling the form and content of the sugar; 2) controlling the temperature; and, 3) stirring correctly. As the solution cools, the sugar crystallizes into the proper size. If the nuclei appear slowly in the syrup solution, there is more time for the sugar molecules to aggregate around the nuclei and form large crystals.

Step 1: Preparation of the sugar solution

This step is basically the same for crystalline and non-crystalline candy. There are different ingredient (formulas) used depending upon the candy recipe. All sugar based candies, whether creamy or chewy or brittle, typically start out with crystalline sugar (sucrose), sometimes along with its close relative such as glucose or corn syrup (invert sugars), as its main ingredient. What determines the type of candy being made is done through the type and proportion of ingredients that make up the initial sugar solution. Weighing ingredients is the most accurate way to measure solids, such as sugar, but it can also be measured in a dry measuring cup. Measuring liquids in a liquid measuring cup or weigh, preparing all equipment and tools in advance; keeping all pots and utensils spotless and dry and work surface is must. If using a buttered pan or mold at the end, always have it ready. Keep a container of ice water handy. Dissolve the sugar (a solute) in liquid, typically water (a solvent): and optionally mixed with other ingredients to create a sugar solution. The sugar and water ingredients are put into pot large enough so boiled sugar does not overflow and placed over medium heat. Stir the mixture constantly until the sugar is dissolved. If one tiny speck of a sugar crystal that hasn't been dissolved falls into the mixture during cooking, the whole batch will return to a solid state, ruining it. Place the sugar in the pot by pouring it in the center. Mix the sugar and water together by drawing an "X" across the sugar, but avoid touching the sides with the sugar. Never stir solution after the sugar crystals are completely dissolved in the water and when it has started boiling - this will incite the formation of big crystals that will make a candy grainy when it cools. Ensuring all crystals are off the side of the pan could be done either oiling the sides or brushing sides of the pan with a heatproof brush dipped in water. Covering the pan during first few minute causes steam forms, condenses and then washes off the side of the pan.

Step 2: Cook (boil) the sugar solution into a concentrated sugar syrup

Sucrose tolerates the high heat of boiling; after a sugar solution is formed, it can be heated and boiled to certain temperatures concentrating the solution as sugar syrup, whereby chemical changes or reactions in the sugar crystals take place. Depending on the candy being made, the syrup is boiled to a codified temperature, measured with a Candy Thermometer, and/or to the syrup's specific concentration indicated on the. Keep the temperature constant; never try to rush a candy mixture by cooking it at a higher temperature than the recipe directs, or slow it down by reducing the heat.

Step 3 and 4: Cooling and beating (optional)
When done, the candy mixture is cooled by pouring its contents onto a marble slab, silpat mat or into a glass bowl or into a pan or mold to harden. Whether cooling and/or stirring the sugar syrup during cooking or afterwards is determined by the type of candy being made. Many of the non-crystalline candies are poured out of pan immediately after cooking. They harden quickly because they are made from highly dehydrated sugar syrups and any agitation or stirring will cause unwanted crystallization. For example, in the case of caramels and hard candy, such as lollipops, their finished cooked recipes are poured directly into their molds or pans, and left to cool.

When boiling stops and the cooling process starts, if everything is done right, the syrup continues to cool as a supersaturated solution and is able to get recrystallized, the size of which is also influenced by stirring, kneading or beating. Fudge is kneaded to break crystals into smaller pieces, making its texture smooth and creamy. As the syrup ahs been stirred, the candy will begin to lose its glossy appearance, becoming somewhat opaque and lighter in color. As crystallization proceeds, the candy will become much thicker and cease to flow. At this point, crystallization is complete and candy should be removed from the pan as quickly as possible and kneaded to dissolve lumps and the mass become creamy and shinny in appearance.

Step 5 : Shaping and final cooling
After kneading, fudge is pressed into a pan and left to set. There are two main ways of forming sweets: cutting into pieces, or setting in molds. Molds may be as simple as greased and lined pans or those which make more complex impressions

Types of Crystalline and Non Crystalline Candies

1. Hard candies and pulled candies:
Hard candies (also called boiled sweets) are single-phase, amorphous sugar candies that are commonly made from a combination of sucrose and glucose syrups. They are typically about 98% or more solid sugar. They have a glassy, translucent appearance. Pulled candy, like rock or Brach's starlight mints, is a hard candy that has been pulled or stretched to incorporate air. This process makes the candy opaque.

2. Fondants: Fondant candy is a partly crystallized, two-phased sugar candy. It is about 88% sugar by weight, usually with much more sucrose than glucose.

In making fondant, a stiff sugar paste is cooked to a high temperature, then carefully cooled and mechanically beaten to produce the desired texture.

3. Caramels and toffees: Caramels contain milk and are cooked to a lower temperature than most sugar candies; toffees are similar, but use less milk and are cooked hotter. In both cases, the milk protein causes these emulsified candies to hold their shapes and prevents the sugars from crystallizing. Their brown color is due to a Maillard reaction between the milk protein and the sugars.

4. Fudges: Fudges, which are made in a wide variety of flavors, are essentially two-phased, crystallized caramels, with a short texture (easily broken). Sugar crystals are formed either due to agitation or the addition of crystal seeds in the form of powdered sugar or crushed fondant candy. The texture depends on the number and size of sugar crystals, the fat content, and the dispersion of milk solids.

5. Nougats and marshmallows: Nougats and marshmallows are confectionery foams, full of air. In the final product, there is often as much air, or even more, than sugar; for marshmallows, a ratio of 5 parts air to two parts syrup by volume is typical. Chemically, they may be single-phase or two-phased. Marshmallows are stabilized by a colloid like gelatin. Compared to nougats, marshmallows have higher moisture content, are softer and more rubbery, and dry out more easily.

6. Jellies and gums: Jellies and gums are thick liquid sugar candies. Gums, such as wine gums, are drier than jellies. They are made from sugar syrup plus a gelling agent. They are cooked to the lowest temperature of all sugar candies and consequently have the highest water content of sugar candies, about 20 to 25% water. Their stiffness depends on the type and amount of gelling agent, the final concentration, the pH of the product, and other factors. The most popular forms of gelling agent are gelatin, agar-agar, starch, and pectin. These produce different effects.

7. Nut pastes: The most common nut paste candy is marzipan, which is an almond nut paste. Nut pastes are made by mixing crushed nuts with sugar syrup. Panned candies: Panned candy is a category of candy that includes dragées and comfits. These candies are formed by coating nuts, preserved fruits, or other candies with either sugar or chocolate in a revolving pan.

8. Pralines, truffles, and noisettes: There is significant variation among pralines, truffles, and noisettes In general, they involve roasting nuts in a high-temperature sugar syrup, and then grinding the cooled result into a paste. Lozenge pastes and cream pastes: Lozenge paste is a sugar candy made by combining fine sugar with a natural gum like gum arabic. The paste is stamped, cut, and dried until almost no water content remains. A cream paste may include gelatin and is not dried as completely

Factors Affecting Crystallization Of Sugars

In making icings, frostings, or candy like fondant and fudge, it is necessary to crystallize the sugar solution. For crystallization to occur, nuclei must form in the solution. To these nuclei the material of the solution is added to form crystals. Both the rate of formation of nuclei and the rate of crystallization are affected by the nature of the crystallizing substance, the concentration, the temperature, agitation, and the impurities present in the solution.

Nature of the crystallizing substance: Some substances like salt crystallize readily from water solution. It requires only a very slight super-saturation to start nuclear formation, and all excess salt in the solution beyond the saturation point is precipitated as crystals. Some substances do not form nuclei or crystallize so readily as salt. With sucrose it is often necessary to have a considerable degree of supersaturation before crystallization commences. Sucrose crystallizes more readily than levulose.

Formation of nuclei: Nuclei cannot form and crystallization cannot occur except from a supersaturated solution. The formation of nuclei, that is the uniting of atoms to form nuclei, is influenced by several factors. If a solution is left to stand, a few nuclei may form spontaneously in various places, and from these nuclei crystallization proceeds. When only a few nuclei develop spontaneously in the solution, the crystals grow to large size. Usually nuclei formation and crystallization do not begin immediately after supersaturation occurs. The rate of nuclear formation may be favored by specks of dust in the solution.

Agitation : Agitation or stirring of a solution increases the rate of nuclear formation. A drop in temperature at first favors, and then retards, the formation of nuclei. Instead of spontaneous formation of nuclei, seeding a solution may be used to start crystallization.

Seeding: When crystals of the same material are added to start crystallization the process is called seeding. These crystals serve as nuclei for crystal growth. If the quantity of crystals added is large and the size of the crystals small, it serves as many nuclei in the solution and the resulting crystals are small. If the quantity of material added is very small, the nuclei formed are few in number and the crystals formed are large. One may think of all crystals as being large enough to be visible, whereas many of them may be very small, so small in fact that they may float in the air. If crystals are floating in the air there is the possibility that they may serve to seed solutions, and thus start crystallization.

Rate of crystallization: To the nuclei formed in the solution new molecules from the solution are deposited, in a regular order or manner, so that each crystal has a typical shape. One side or face of a crystal may grow more rapidly than another. The rate at which the nuclei grow to larger size is called the rate of crystallization. This rate may be favored by the concentration of the solution and its temperature; it may be hindered by foreign substances.

Concentration of the solution: A more concentrated solution favors the formation of nuclei. Fondant syrup cooked to 114°C. Contains less water and is more concentrated than one cooked to 111°C. Thus nuclei form more readily in the one cooked to 114°C. Large, wellshaped crystals form more readily if the degree of supersaturation is not too great. The most favorable supersaturation for crystal growth, of a sucrose solution boiled to 112°C, is that between 70° and 90°C. Although crystallization occurs in a very short time when the syrup is stirred at these temperatures, the crystals formed are larger than when the syrup is cooled to a lower temperature. Supersaturation and a low temperature are desirable for the development of small crystals.

Temperature at which crystallization occurs: It is a well-known fact that, in general, chemical precipitates come down more coarsely crystalline if crystallized at high temperatures. The sugars follow this general rule. Other things being equal, i.e., concentration, etc., the higher the temperature at which crystal formation occurs, the coarser the crystals formed. A drop in temperature at first favors the formation of nuclei, and then hinders it. Crystallization is favored in sugar syrups by cooling to a certain temperature, but is hindered when cooled to a lower temperature. Since the viscosity of a saturated sugar solution becomes increasingly greater as the temperature falls below 70°C, crystal formation is also slower as the temperature falls.

Agitation: Stirring a solution favors the formation of nuclei and hinders the depositing of the material of the solution on the nuclei already formed. Hence, crystals in solutions that are stirred do not develop to the size that they do in spontaneous crystallization. If small crystals are desired, then the conditions should be such that many nuclei are formed. Small crystals are obtained in syrups of definite concentration and temperature, if the syrup is stirred until the mass is kneadable. However, if the syrup is stirred for only a short time, some nuclei are formed, but after agitation is stopped, the formation of new nuclei is not favored and crystal growth is favored. This emphasizes the importance of stirring candy and icing syrups until practically all the material is crystallized, if small crystals are desired. Impurities in the sugar syrup may also result in the formation large sugar crystals. Impurities promote premature crystal formation, which will grow to big unwanted crystals. Interfering substances: Some products can be added prevent the formation and growth of crystals. These products such as cream, butter, egg white etc., are called interfering agents. The agents coat the crystals and prevent the growth of large crystals. Boiling the sugar syrup to the exact temperature is also very important, complete solution of the sugar is very important.

Degree of inversion: Sweets containing high concentrations of sugar (sucrose) may crystallize either during manufacture or on storage (commonly referred to as graining). Although this may be desirable for certain products (such as fondant and fudge), in most other cases it is seen as a quality defect. When a sugar solution is heated a certain percentage of sucrose breaks down in to 'invert sugar'. This invert sugar inhibits sucrose crystallization and increases the overall concentration of sugars in the mixture. This natural process of inversion, however, makes it difficult to accurately assess the degree of invert sugar that will be produced. As a way of controlling the amount of inversion, certain ingredients, such as cream of tartar or citric acid, may be used. Such ingredients accelerate the breakdown of sucrose into invert sugar, and thereby increase the overall percentage of invert sugar in the solution. A more accurate method of ensuring the correct balance of invert sugar is to add glucose syrup, as this will directly increase the proportion of invert sugar in the mixture. The amount of invert sugar in the sweet must be controlled, as too much may make the sweet prone to take up water from the air and become sticky. Too little invert sugar will be insufficient to prevent crystallization of the sucrose. About 10–15 per cent of invert sugar is the amount required to give a non-crystalline product.

Added ingredients: The addition of certain ingredients can affect the temperature of boiling. For example, if liquid milk is used in the production of toffees, the moisture content of the mixture immediately increases, and will therefore require a longer boiling time in order to reach the desired moisture content. Added ingredients also have an effect on the shelf-life of the sweet. Toffees, caramels, and fudges, which contain milk-solids and fat, have a higher viscosity, which controls crystallization. On the other hand, the use of fats may make the sweet prone to rancidity, and consequently the shelf-life will be shortened.

Chewing Gums

Chewing gum is a sweetened, flavored confection composed primarily of latex, both natural and artificial. Organic latex, a milky white fluid produced by a variety of seed plants, is best known as the principle component of rubber. Used as a snack, gum has no nutritive value, and, when people have finished chewing, they generally throw it away rather than swallow it.

The manufacture of chewing gum in the United States has come a long way from loggers chopping off wads of spruce gum for chewing pleasure, yet the base of the gum remains the sap of various rubber trees, or, in most cases, a synthetic substitute for such sap. Natural gum bases include latexes like chicle, jelutong, gutta-percha, and pine rosin. Increasingly, natural resins other than chicle have been used because chicle is in extremely short supply: a chicle tree yields only 35 ounces (one kilogram) of chicle every three to four years, and no chicle plantations were ever established. However, natural latex in general is being replaced by synthetic substitutes. Most modern chewing gum bases use either no natural rubber at all, or a minimal amount ranging from ten to twenty percent, with synthetic rubbers such as butadiene-styrene rubber, polyethylene, and polyvinyl acetate making up the rest.

After the latex used to form bases, the most common ingredient in chewing gum is some type of sweetener. A typical stick contains 79 percent sugar or artificial sweetener. Natural sugars include cane sugar, corn syrup, or dextrose, and artificial sweeteners can be saccharine or aspartame. Popular mint flavors such as spearmint and peppermint are usually provided by oils extracted from only the best, most aromatic plants. Thus, while the aroma of a stick of spearmint gum is quite strong, flavoring comprises only one percent of the gum›s total weight. Fruit flavors generally derive from artificial flavorings, because the amount of fruit grown cannot meet the demand. For example, apple flavor comes from ethyl acetate, and cherry from benzaldehyde. In addition to sweeteners and flavorings, preservatives such as butylated hydroxytoluene and softeners like refined vegetable oil are added to keep the gum fresh, soft, and moist. Fillers such as calcium carbonate and corn starch are also common.

Federal regulations allow a typical list of ingredients on a pack of chewing gum to read like this: gum base, sugar, corn syrup, natural and/or artificial flavor, softeners, and BHT (added to preserve freshness). This vagueness is mainly due to the chewing gum manufacturers' insistence that all materials used are part of a trade secret formula.

Manufacturing Process

While the specific ingredients in gum might be a secret, the process for making gum is not. The first chewing gum making machine wasn't even patented, and today the procedure is considered standard throughout the industry.

Preparing the chicle

If natural latex is to be used, it must first be harvested and processed. The tall 32.79 yard (30-meter) chicle tree is scored with a series of shallow Xs, enabling the chicle to flow down into a bucket. After a significant amount of chicle has accumulated, it is strained and placed in large kettles. Stirred constantly, it is boiled until it reduces to two-thirds of its original volume. It is then poured into greased wooden molds and shipped.

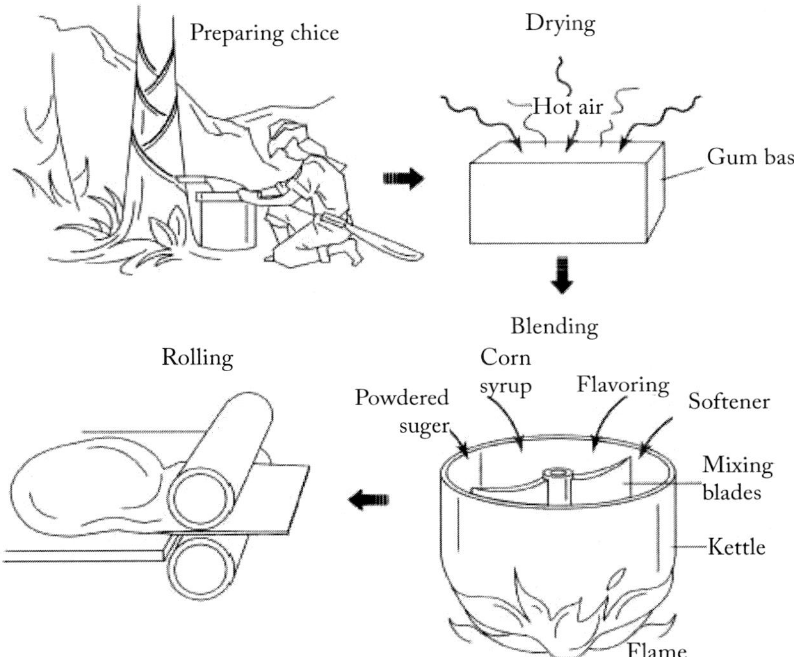

Chewing gum base consists either of natural latex or a synthetic substitute. Natural latex such as chicle is harvested by making large X-marks on rubber trees and then collecting the substance as it runs down the tree. After grinding the base to form a coarse meal, the mixture is dryed for a day or two. Next, the mixture is heated in large kettles while the other ingredients are added. Large machines then pummel, or "knead," the mass until it is properly smooth and rubbery, and it is put on a rolling slab and reduced to the proper thickness.

Kneading and rolling the gum

The next step is kneading. For several hours machines gently pummel the mass of chewing gum until it is properly rubbery and smooth. Large chunks are then chopped off the mass, to be flattened by rollers until they reach the proper thickness of nearly .17 inches (about .43 cm). During this process, the sheet of chewing gum is dusted with powdered sugar to prepare it for cutting.

Cutting and seasoning the gum

A cutting machine first scores the sheet in a pattern of rectangles, each 1.3 inches (3.3 centimeters) long and .449 of an inch (1.14 centimeters) wide. The sheet is then put aside at the proper temperature and humidity to "season."

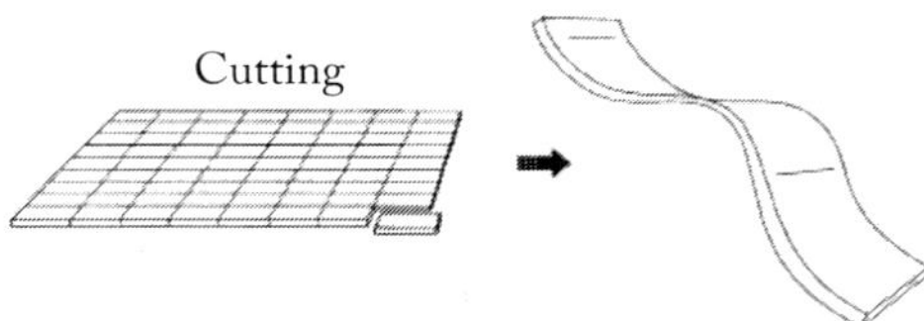

After being dusted with powdered sugar, the gum is scored into a pattern of rectangles, seasoned, and broken into sticks. The gum is now ready to be packaged and shipped to retail outlets.

Packaging the gum

Once seasoned, the gum sheets are broken into sticks, wrapped in aluminum foil or wax paper, wrapped in paper, and put into plastic packs that are then sealed.

References

Kennedy, Angus (2008). Kennedy's Confection Magazine.

Peter Koch. The Enrobing Process and techniques. The manufacturing Confectioner, June 2011.

Richardson, Tim H. (2002). Sweets: A History of Candy. Bloomsbury USA. ISBN 1-58234-229-6.

Weatherley, Henry (1865). A Treatise on the Art of Boiling Sugar. Retrieved 2008-07-14

13
Equipments Used in Confectionery Manufacturing

Chocolate is a product that requires complex procedures to produce. The process involves harvesting coca, refining coca to cocoa beans, and shipping the cocoa beans to the manufacturing factory for cleaning, roasting, grinding , refining , conching , moulding and enrobing operations which are accomplished using wide range of machinery. A few have been discussed in this chapter.

1. Bean Roasters

When cocoa beans were first brought to Europe, the method of roasting was quite simple, often consisting of no more than metal trays suspended by chains over a fire, an ancient Middle Eastern technique for roasting coffee. The beans would be stirred by hand with a long stick or paddle until the roast was complete. Needless to day, this was very labor intensive and inefficient. Clearly, something more efficient was needed. The next development in roasting technology was not that far distant from roasters in use today. Metal cylinders or balls that could hold up to several pounds of cocoa beans would be suspended over the embers of a fire. A rod would run through the center and act as an axis upon which the roaster would turn. A hand crank on one end would allow the cylinder or ball to be turned while the beans were roasting. The result was a much more even roast than was before possible. Aspects of this technique remain in many of today's roasters.

Ball Roasters

As mentioned, all roasters were one of the very first kinds of roaster: balls were simply suspended over a fire, the cocoa beans inside. Ball roasters remained one of the primary and traditional roasters for cocoa beans through the turn of the century and were constructed at least through the 1960's. Later designs were pioneered by the company Sirocco. (Sirocco is named after the hot wind that travels up from the Sahara desert and heads north and may reach hurricane speeds by the time it reaches northern Africa, Spain and France.)

The ball is encircled within a secondary chamber so that the hot air passes in between the ball and the outer chamber wall. This prevents the flames from impinging directly on the ball and makes it easier to over roast the beans.

Today, ball roasters have been largely replaced by roasters of similar design that use a rotating cylinder rather than a rotating ball. There are a number of practical advantages of this design, since the shape of a cylinder with two ends makes it easier for samples to be removed. It also allows for easier measurement of the temperature of the beans if automation is a requirement. Ball roasters still hold advantage when flavor is of paramount importance.

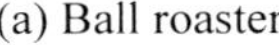
(a) Ball roaster

(b) Roasting in a cylinder

Fluid Bed Roasters

Fluid bed roasters are a relatively new development. There are a number of different designs, some of them radically different. What they all have in common is that air is blown up through the beans, suspending them at least temporarily in air. The bottom of the roaster may vibrate to help launch the beans into the air. Fluid-bed roasters are typically used in large industrial environments, since they are conducive to short, high-temperature roasting that allows large volumes of beans to be roasted in a short period of time.

Continuous Roasters

The large chocolate manufacturers rely on continuous roasters. Given the quantity of beans that must be roasted for a large industrial chocolate manufacturer and the relatively low prices of industrial chocolate, they are probably one of the only economical ways to roast the large quantity of beans that must be roasted. There are many different styles of continuous roaster, though in general they fall into two different categories. In one style, the beans are passed on a conveyer belt through a roasting tunnel. The degree of roast can be controlled by controlling the temperature in the tunnel as well as the speed of the conveyer belt. The heat is applied to the beans in a consistent way throughout the roasting process. Alternatively, another style uses a tall tower into which the beans are fed into the top. The tower has a series of trays upon which the beans rest. Once the beans have roasted on one tray for a given period of time, they are dumped onto the next tray. By the time the beans reach the bottom of the tower, they are fully roasted. The airflow through the tower is from the bottom up. The beans cool the air as the air passes them, at the same time heating the beans. This means that the beans on the lower levels are heated faster than those at the top. Because of this, they will generate a different flavor profile in the finished chocolate. It should be pointed out that there are also continuous fluid-bed roasters.

One disadvantage of this kind of roaster is that it is relatively difficult to pull samples of the beans from the roaster while they are being roasted. Furthermore, if the beans are found to be properly roasted when they are only partially through the roaster, there is no way to quickly remove them from the roaster and get them into a cooler. On the other hand, the large industrial chocolate makers run such large quantities of beans, this process can be fine tuned for a particular type of bean. Furthermore, most of these manufacturers use Forestero beans almost exclusively, for which the degree of roast is not as critical as it is with Criollo and other types of high-grade flavor beans.

Continuous Cocoa bean roaster

Cocoa Mass Roasters

Cocoa mass roasters are basically tanks in which the ground cocoa bean paste (cocoa mass) is heated as it is stirred until the desired level of roast is reached. It may also be implemented as a large heated cylinder where the cocoa mass is spread in a thin film. Since the film is thin, it roasts very quickly, and a thin knife blade scrapes it off of the far side of the cylinder. Cocoa-mass roasters are used by the large industrial chocolate manufacturers as a way to quickly roast large quantities of cocoa beans with minimal labor, since it is also a process that by nature requires little labor, and it also lends itself to easy automation.

There are even combinations of the various kinds, such as continuous fluid bed roasters. No matter the technology, the best roaster is the one best suited to the job. With poor quality beans, cocoa mass roasters and continuous roasters, or even coffee roasters, will often suffice, since the beans' flavors are already considerably damaged; no matter the care, the lost flavors may not be brought back. However, if fine flavor is to be developed in the finished chocolate, the choice of roaster is much more critical, as are the techniques used to roast the cocoa beans. Different beans and different styles of chocolate have different roasting requirements, so care must be taken to select the roaster that will best bring the final result, be it high production quantity, as is the case for the large industrial manufacturers, or fine flavor for the artisanal chocolate makers.

Roasting truly is one of the chocolate maker's arts. It takes a fine flavor palate to be able to judge a roast and judge how the roasted cocoa bean will taste when it is turned into fine chocolate. An incredible number of flavor changes occur during the time the cocoa beans are roasted. Even after the chocolate is complete—even as the chocolate is "resting" and undergoing its final stages of flavor development—the flavors continue to change. The resting process continues oftentimes for several months. The chocolate maker who seeks for optimal flavor must not only think about the flavor of the bean as it is being roasted but must think about how the flavors will change as the chocolate is being made and how the chocolate will taste after it has rested. The flavors will be radically different, and new flavor profiles will often express themselves. Roasting is a bit of a secret art. There are no guidelines on how the best cocoa roasts are to be obtained. The roaster is an artist who uses one of the world's greatest sources of flavor and noblest of ingredients—the cocoa bean, also known as "the food of the gods."

2. Nib Grinder

Grinding cocoa nib into liquor prior to refining improves the refining process. The Nib Grinder is an impact hammer mill, supplied with three grinding screens. The hammers and screens are hardened by a patented process to improve wear and minimize metal in the product. The machine has a variable speed drive, controlled by a keypad on the control panel. The Grinder infeed is guarded, and there is an E-Stop on the control panel for safety.

The grinding screens have different size holes, and are easily interchangeable. Winnowed nib is first fed through the grinding screen with the largest hole openings. The coarse powder from the first grind can then be fed through the grinding screen with the medium hole openings, resulting in a liquor paste, as fat begins to release from the nib. A third pass can be made through the grinding screen with the small hole openings, yielding a fluid liquor typically below 200 microns.

Grinding the nib in stages reduces the heat impact on the liquor from the grinding process. There are faster methods of grinding nib, but the resulting liquor may also experience much higher temperatures from the shear. Perfectly roasted nib are never exposed to high temperatures in the grinding process

Cocoa nib grinder

Nib powder

Grinding the nib in stages reduces the heat impact on the liquor from the grinding process. There are faster methods of grinding nib, but the resulting liquor may also experience much higher temperatures from the shear. We don't want to expose our perfectly roasted nib to high temperatures in the grinding process

3. 5 Roll Chocolate Refiner

Refining is the final grinding of all particles in the liquid chocolate together to produce an even extremely smooth texture in which no grit can be detected on one›s tongue or pallet.

5. 5 ROLL REFINER for chocolate is designed with the latest technology. 5 roll refiner can be used as finer after 2 roll refiner or stand alone for refining process.

It is possible to get high degrees of fineness from 5 five roll refiner without causing damage to the product due to the overheating which is typical of other types of refining processes. We obtained very fine outputs during our tests.

The area of contact between the rolls is very limited and 50% of the surface of the rolls is used as a cooling area. 5Roll Refiner with 50cm rolls are used for getting fine cocoa mass.

The results obtainable with the use of the 5 five roll refiner in particular, the fineness, flavor and technological characteristics of the finished product cannot be reproduced with other types of systems. This is because other processes take place during the refining process, and these are fundamental for the formation of the characteristics of the final product.

Here the particles are ground to their final size, usually below 30μm in order to avoid a sandy texture in the mouth in the final product.

5 Roll Refiner

4. Conche

A conche is a surface scraping mixer and agitator that evenly distributes cocoa butter within chocolate, and may act as a "polisher" of the particles. [It also promotes flavor development through frictional heat, release of volatiles and acids, and oxidation. There are numerous designs of conches. Food scientists are still studying precisely what happens during conching and why. The name arises from the shape of the vessels initially used which resembled conch shells.

When ingredients are mixed in this way, sometimes for up to 78 hours,chocolate can be produced with a mild, rich taste. Lower quality chocolate is conched for as little as six hours. Since the process is so important to the final texture and flavor of chocolate, manufacturers keep the details of their conching process proprietary. Rodolphe Lindt invented the "conche" in Berne, Switzerland in 1879. It produced chocolate with superior aroma and melting characteristics compared to other processes used at that time. Legend has it that Lindt mistakenly left a mixer containing chocolate running overnight. Though he was initially distraught at the waste of energyand machine wear and tear, he quickly realized he had made a major breakthrough. Before conching was invented, solid chocolate was gritty and not very popular. Lindt's invention rapidly changed chocolate from being mainly a drink to being made into barsand other confections.

Lindt's original conche consisted of a granite roller and granite trough; such a configuration is now called a "long conche" and can take more than a day to process a tonne of chocolate. The ends of the trough were shaped to allow the chocolate to be thrown back over the roller at the end of each stroke, increasing the surface area exposed to air. A modern rotary conche can process 3–10 tonnes of chocolate in less than 12 hours. Modern conches have cooled jacketed vessels containing long mixer shafts with radial arms that press the chocolate against vessel sides. A single machine can carry out all the steps of grinding, mixing, and conching required for small batches of chocolate.

Conching redistributes the substances from the dry cocoa that create flavor into the fat phase. Air flowing through the conche removes some unwanted acetic, propionic, and butyric acids from the chocolate and reduces moisture. A small amount of moisture greatly increases viscosity of the finished chocolate so machinery is cleaned with cocoa butter instead of water. Some of the substances produced in roasting of cocoa beans are oxidized in the conche, mellowing the flavor of the product.The temperature of the conche is controlled and varies for different types of chocolate. Generally higher temperature leads to a shorter required processing time. Temperature varies from around 49 °C (120 °F) for milk chocolate to up to 82 °C (180 °F) for dark chocolate. The elevated temperature leads to a partially caramelized flavor and in milk chocolate promotes the Maillard reaction.

The chocolate passes through three phases during conching. In the dry phase the material is in powdery form, and the mixing coats the particles with fat. Air movement through the conche removes some moisture and volatile substances, which may give an acidic note to the flavor. Moisture balance affects the flavor and texture of the finished product because, after the particles are coated with fat, moisture and volatile chemicals are less likely to escape. In the pasty phase more of the particles are coated with the fats from the cocoa. The power required to turn the conche shafts increases at this step. The final liquid phase allows minor adjustment to the viscosity of the finished product by addition of fats and emulsifiers, depending on the intended use of the chocolate.

While most conches are batch process machines, continuous flow conches separate the stages with weirs over which the product travels through separate parts of the machine. A continuous conche can reduce the conching time for milk chocolate to as little as four hours.

Macintyre system

This very unique machine which resurrects the traditional method of conching and grinding at the same time. It consists of a double-jacket cylinder with serrated internal surface. Spring-loaded scrapers break the particles during rotation; volatile water and flavours are removed by ventilation and heating. There has been optimisation of flow properties and flavour in those machines and it has also been tried to combine it with other systems, e.g. refiners. It is also known, that operation is relatively noisy. An advantage is that batch sizes between 45kg and 5t are possible, which means a lot of flexibility for smaller companies.

5. Enrobers in the Confectionery Industry

Enrobing is a process that involves covering a confection or snack with chocolate or chocolate coatings. Traditionally, this process was slow and involved manually dipping the pieces into melted chocolate by hand. As demand for chocolate-coated sweets increased, it became impractical or impossible to employ enough people to dip sweets into melted chocolate to keep up with production demand. Enrobing can be carried out with chocolate or compound coatings (compound coating is a replacement product made from a combination of cocoa, vegetable fat, and sweeteners). An advantage of compound coatings is that they may set faster and no tempering (the process in which chocolate masses are thermally treated to produce a small fraction of homogeneously dispersed, highly stable fat crystals of the correct type and size) is needed[1]. Some typical examples of enrobed products are wafer bars, fondant centres, jellies, nuts, biscuits and ice cream.

Through covering the centre with chocolate or compound coatings, the shelf-life of the product may be extended. This is primarily applicable to centres that, if not covered, could be prone to moisture uptake/loss, oxidation, or microbial spoilage. Enrobing has some advantages over moulding (which is another method of getting a chocolate covered centre) such as greater production rates, lower capital costs. Enrobing often allow for greater production rates with lower capital costs than moulding.

Enrobing Process

Enrobers are available in different sizes, suitable for large and small scale production and there is a wide variety of different designs to meet all requirements. Belt widths from 125 millimetres to 2600 millimetres are available. Although the basic elements of an enrober have largely remained unchanged over the years, the methods of operation and degree of precision possible have changed significantly. This has been accompanied by a modest increase in throughput. The biggest change in the manufacturing of chocolate enrobed sweets can be credited to the efficiency of the coolers.

Processing a real chocolate always requires a tempering unit. Whether fitted with a temperer internally or externally, enrobers have the same basic components. It is important that the centres entering the enrober are maintained at 21–24°C, and that the enrobing chocolate has the desired viscosity and rheological properties. Warmer centres may lead to possible bloom problems due to the residual heat increasing the chocolate temperature of the enrobed sweets. Cold centres can lead to blooming and cracking of the coating shell due to expansion of the centre mass as it warms. Fat bloom develops in different ways. Automatic crystal conversion on the one hand caused by incorrect and/or insufficient tempering; on the other, it may be caused by fat migration from the filling where this fat penetrates the chocolate coat and causes the cocoa butter crystals to rise through the surface. Loss of temper can also be due to heat damage.

The centres are fed on to a feed band and transferred to a wire belt, which passes through the enrober. The coating medium is maintained at a constant temperature and in a controlled condition in an agitated tank; it is then pumped to a flow pan. The flow pan aids the process by creating a continuous curtain of coating and feeding a bottoming device. This leads to the formation of a bed of coating, which floods the mesh band. The centres are passed through this curtain and bed and are covered on all surfaces. After the curtain, excess chocolate is forced off the product by an air blower and a licking roller is used to control the amount of mass left on underside of the sweet. There is normally a vibrator after the blower to remove excess chocolate and to improve the appearance of the sweet. Finally, there should be a detailing rod between and end of the wire belt and the start of the cooler belt. Following the curtain, using an air blower, the excess chocolate is removed and a licking roller controls the amount of mass left on the underside of the sweets.

The excess mass from the curtain falls through the wire mesh belt into a sump, and is recirculated. Part of the mass is diverted through a de-temperer and is then re-tempered; blending of the freshly tempered and recirculated streams controls the overall level of chocolate in the enrober. Vibrating the centres removes ripples left by the fan, smoothens the coating and removes any surplus chocolate returning it to the tank. The centres are discharged from the enrober on to a cooling conveyor passing over a de-tailer which is a rapidly spinning rod across the end of the wire band. This results in the removal of the tail that forms as the centre leaves the wire band.

Cooling

After enrobing, the product enters a cooling tunnel to allow the coating to harden. To avoid blooming problems, temperature changes in the tunnel should be gradual, and the relative humidity properly controlled. If the dew point is lower than room temperature, moisture could condense on the product and cause sugar bloom during storage[4]. The chocolate coating and the filling are cooled down to approximately 18°C to ensure trouble-free packaging. A good cooling tunnel should be divided into three zones.

Convection chocolate cooling is a time-dependent but not energy-dependent process. A higher temperature and longer cooling time are more favourable than a lower temperature and shorter cooling time. The recommended cooling times for pure dark chocolate, milk chocolate and milk chocolate with CBE portions are approximately six, eight and 12 minutes respectively. Milk chocolate requires a longer cooling time than dark chocolate due to the higher milk fat content and consequent lower solidification temperatures. Compound coatings may require different cooling profiles i.e. adjusted to suit the setting properties of the vegetable fat used. After cooling, a product which is shiny and resistant to handling should be available for packaging.

Chocolate enrobing

Enrobed milk chocolate

6. Counter-rotating Twin-screw Extruders

Counter-rotating twin-screw extruders are widely used in a wide range of product manufacture in chocolate and confectionery industries. Reliability, flexibility and high throughput are the most important demands placed on a solution for the extrusion processes. Two or more extruders can be easily combined to produce co-extruded products. A wide range of extruding heads and nozzles are available, thus allowing for a great variety of different combinations of colors, flavors and shapes. Besides, co-extruded products can be also center-filled. The extruders are ideally suited for the production of centre filled products. The high pressure stability of the screws provides a constant an even flow of the filling mass (bubble- gum, chewy candy, marzipan, etc). Also, additional filling devices can be attached to the extruding head for the production of semi-liquid filled products.

Confectionery Processing uses Twin Screw Extruder for manufacturing Chewing Gum. Twin screw extruders are used in the continuous production of chewing gum to replace the more traditional batch methods which use large, labor intensive and high energy kneaders. There are two main fields of application, first compounding of the gum base by extrusion, and secondly processing the final chewing/ bubble gum mass. For the latter, ingredients such as gum base and various additives are fed into the infeed section of the extruder using high accuracy feeders. The extruder plasticizes the mass by means of high shear forces. Subsequently, sugar, sugar-like substances, sweeteners, flavors and other additives are added by means of a feed assembly which consists of a feeder above a twin screw side feeder. The ingredients are fed directly into the plasticized mass in the process section. The extruder screws homogeneously disperse all ingredients into the gum matrix, ensuring best flavor retention and optimal release of flavor when the product is consumed

Extrusion of Chocolate Crumb

Twin screw extruders are in use for several different applications in chocolate manufacturing. For crumb chocolate flavor development, sugar and milk powder are fed into the extruder via highly accurate feeders. Liquid components like water and cocoa mass are metered in by feeding pumps as well. Inside the extruder, the screws generate the mass to high shear forces with resultant high temperatures. As a result the Maillard reaction for creation of chocolate crumb flavor takes place. Undesired volatiles and off-flavors native to cocoa are removed by atmospheric or vacuum degassing utilizing the twin screw side devolatilization unit. At the end of the extruder, the crumb is pelletized by the GF centric food pelletizer. A further field of application is melting of chocolate powder coming from refiner rolls and mixing it with other ingredients like lecithin and cocoa butter. In this process step, the rheological properties of the chocolate melt can be adjusted by shear applied inside the extruder.

Extrusion of Caramel Masses

Another field of application in the confectionery industry is the extrusion of caramel masses. Sugar and maltodextrins are fed via highly accurate feeders into the extruder. Inside the process section of the extruder the ingredients are melted by means of shear forces and mixed with milk powders containing protein. In a High Temperature Short Time (HTST) process, the Maillard reaction takes place creating caramel flavor. Fat, nuts, flavors or other ingredients may subsequently be fed into the extruder by means of a feed assembly, which consists of a feeder above a twin screw side feeder, and gently mixed into the caramel mass. The product is often then discharged to a cooler belt to immediately stop the Maillard reaction.

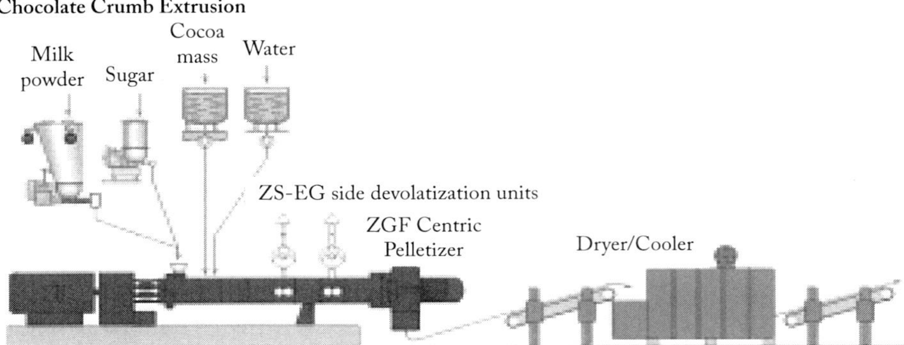

7. Mesh Extrusion

Mesh-extrusion is a fascinating variety to block-type chocolate. Lots of fine chocolate strings are extruded as a web on a continuously running transport belt. The woven mesh layers can consist of different colors which of course increase the visual appeal of the product. Mesh-extrusion consists of several extruding heads which are located one after the other over a transport belt.

A diagonal pattern of thin chocolate strings one beneath the other is obtained by the movement of the extrusion head back and forth across the belt width, together with the movement of the transport belt itself. The second, third etc. layer is made by another extrusion head. Therefore the different layers look like a woven structure. The number of layers and the diameter of the strings determine the height of the product. With several extrusion heads it makes sense to work with different chocolate colors. Another possibility is to divide the hopper of one extrusion head into different sections and also work with different chocolate colors.

References

Chocolate, Cocoa and Confectionery: Science and Technology By Bernard W. Minifie

http://www.fao.org/wairdocs/x5434e/x5434e0a.htm

14

Packaging and Storage of Bakery and Confectionery Products

India's bakery market at ₹49.5 billion tonnes makes it the third largest market in Asia Pacific, only after Japan and Australia. 72 The bakery industry in India comprises of organized and unorganized sectors. The organized sector consists of large, medium and small-scale manufacturers who produce packaged biscuits and bread. The unorganized sector consists of small bakery units, cottage and household type manufacturing their goods without much packaging and distributing their goods in the surrounding areas. Bread market is estimated to be growing at around 7% p.a. in volume terms, whereas the biscuit market has witnessed a higher growth at around 8–10%. Within the biscuit category, cream and specialty biscuits are growing at a faster rate of 20% p.a. The per-capita consumption of biscuits in India is around 900 gm as compared to 15–20 kg. for developed countries. The consumption of biscuits is equally divided between the urban and rural population. Demand for biscuits in 2003–2004 is likely to exceed 1.2 million tonnes.

Packaging of Bakery Products

The packaging of bakery products is closely interlinked with production, preservation, storage, transportation and marketing. The importance of packaging can further be gauged from the fact that packaging constitutes a fair portion (10–25%) of the entire cost of the pack. Product Range Bakery products contain high nutritive value and are manufactured from wheat-flour, sugar, baking powder, condensed milk, ghee (fat), salt, jelly, dry fruits, various essences and flavouring etc. Different type of bakery products can be classified as:

Dry Bakery Products

- **Biscuits:** Soft biscuits, hard biscuits, cookies, crackers, fancy biscuits, cream wafer biscuits.

Moist Bakery Products

- **Bread:** Sweet bread, Milk bread, Masala bread, Garlic bread, Fruit bread etc.
- **Buns:** Fruit buns, hamburger buns, dinner rolls, crisp bread, pizza.
- **Others:** Cakes, pastries, doughnuts, muffins etc.

Product Characteristics

Dry Bakery Products: These products are fragile and characterized by a low moisture content (<6%) low water activity (Aw = 0.30) and are highly hygroscopic. Moisture is the decisive criteria for the organoleptic properties and acceptability by the consumers. The basic characteristics of dry bakery products are given in Table 1.

Table 1 Basic Characteristics of Dry Bakery Products

Properties	Basic Characteristics of Products
General	Foodstuffs for long storage
Physical-mechanical	Fragile Light Low resistance to moisture Variable sizes
Organoleptic	Crisp or crunchy texture Distinctive flavour Flavors that may change (loss of initial flavour or ingress of foreign flavour) Flavour that may deteriorate (go stale, soapy or bitter, etc.)
Physics-chemical	Low moisture content Hygroscopic Containing fatty matter Greasy surface Sensitive to: • Oxidation • Enzymatic reactions • Non-enzymatic browning • Light
Technical-economical	Industrial Low sales price

Loss of Crispness: Biscuits have a low moisture content, high fat level and are fragile in nature. Hence, they have to be protected from these three aspects. Since the biscuits consists of wheat flour, fat and shortening, sugar, salt and flavouring agents they are pre-dominantly sensitive to water vapour interchanges (moisture) and oxygen reactions. They generally have an initial moisture content of 2–3% equilibrating to 10–15% RH. The critical moisture level from the point of loss of crispness varies between 4 to 6%. The shelf-life of biscuits depends upon:

- Inherent characteristics of the product
- Barrier and other functional properties of the packaging material
- Packaging operations adopted
- Distribution and storage patterns followed
- Economic considerations

There is a well-established relationship between water vapour sorptions and chemical, physical and stability characteristics of biscuits. For predicting product shelf-life and package performance with respect to water vapour transfer, the data required is:

- Water sorption isotherm
- Water Vapour Transmission Rate (WVTR) of the packaging material
- Storage Conditions

Since these are moisture sensitive products, water vapour transmission rate of the packaging material used is of importance as it is closely associated with drying, physical structure and protective action against oxidation. These products not only become brittle and hard but also develop oxidative rancidity at very low moisture contents. Temperature also plays a very important role. As the temperature increases the critical moisture gets reduced since the Equilibrium Moisture Content (EMC) corresponding to the same water activity is decreased. The sensitivity of three types of biscuits is given in Figure 1.

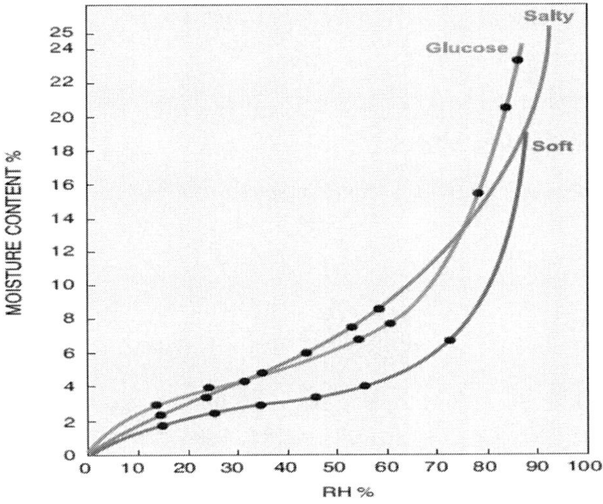

Fig 1 Moisture Sorption Isotherm of Biscuits

IS 1011-1992 specifies the maximum limits for moisture, acidity and acid insoluble ash for biscuits. Table 2 gives the requirements for biscuits.

Table 2 IS 1011-1992 Requirement for Biscuits

Characteristics	Requirement
Moisture, percent by weight, maximum	5.00
Acid insoluble ash (on dry basis), %	0.05

Rancidity: Another requirement due to high fat is the prevention of rancidity. When fat gets exposed to moisture and atmosphere, it gets oxidized and this results in rancidity and lowering of shelf-life. Fruits and nuts used are also susceptible to oxidation in presence of oxygen. Hence the packaging material must be grease resistant to prevent seepage of fat and staining of the pack and have low oxygen permeability to prevent oxidation and rancidity of the fat. Light is also detrimental to colours or cause oxidation of fats leading to rancidity producing undesirable off-flavours. In such cases, opaque packaging material is used. Some biscuits are susceptible to tainting by inks, adhesives and coatings used in the packaging material. The packaging material, therefore, should be free of residual solvents etc; to avoid development of off-flavours.

There are several basic requirements of a package intended to contain bakery products.

These include:

- Water vapour permeability of packages
- Oxygen exchange from within and outside a package
- Aroma impermeability characteristics of packaging materials
- Resistance to seepage of fats and oils
- Protection against deteriorative visible and ultra violate radiation
- Good printability and appearance
- Physical, mechanical protection to the products against shocks, crushing and vibrations
- Compatibility and safety of the packages
- Packaging, in general, must meet the following four basic requirements:
- The biscuit pack must give mechanical protection to the product.

This can be achieved either by packing the biscuits in end-fold style portion packs or by gas flushing the pillow packs, thus preventing breakage during transport and retail handling. Ready to sell individually wrapped packs eliminate the hygiene factor risks since the biscuits do not come in contact with the external environment. The appropriate films and correct sealing prevents any infestation by insects. The result is a product, which is fresh and tasty throughout its shelf-life. The pack must be perfectly heat-sealed in wrapping materials with the required barrier properties against light, humidity and external odours.

- The packaging must appeal to the potential customers and stand out against other competing products and serve as an effective advertising tool.
- Detailed information about the product such as composition of the product, nutritional value, price information etc.
- Satisfy consumer demand for convenience packaging by providing different pack sizes, convenient packet opening facilities like tear tapes, incision cuts etc.

Moist Bakery Products

Breads and cakes are another category of baking products with comparatively less shelf-life. These products have high moisture content (>12%), supple texture and high water activity between 0.6 to 0.85 with low resistance and tendency to crumble and go stale. Their basic characteristics are given in Table 3

Table 3 Basic Characteristics of Moist Bakery Products

Properties	Basic Characteristics of Products
General	Foodstuffs for long storage
Physical-mechanical	Fragile Light Low resistance to moisture Varied sizes and shapes
Organoleptic	Supple and creamy texture Distinctive flavour Flavors that may change (loss of initial flavour or fixation of foreign orders) Flavour that may deteriorate (go stale, soapy) Appearance that may change (drying out)
Physics-chemical	High moisture content High Aw Sensitive to: • Oxidation • Enzymatic reactions • Microbioogical alteration

Since it contains hydrated starch it is prone to staling, thus limiting its shelf-life. It has low fat content and short distribution life, hence it does not need protection against oxygen. Bread is also susceptible to loss in aroma/flavour, so

Plastic Packages of Cakes

Cakes Packed in Plastic Pouches and Trays

the packaging material used must prevent pick up of undesirable off-flavours. The undesirable changes during storage include moisture loss, staling and loss of freshness. The packaging material must possess moderately effective moisture barrier properties. The inner portion of bread has equilibrium humidity in the range of 90%, hence it tends to dry out rapidly and becomes harder. The crust however, has low equilibrium humidity and it tends to become soggy under moist conditions. Too good a moisture barrier, has effect of promoting mold growth on the bread and allows the bread to become soft. If a poor barrier film is used, the bread will tend to dry out and stale. Staling of bread starts with in 3–4 days of manufacturing. This is an inherent property of the type of flour, method of baking and storage conditions. It is caused by the migration of water from the starch to the protein portion of the interior, the starch then becomes dry and looses texture. Since this activity is independent of the moisture content of the inner portion of the bread, an effective packaging material must protect the bread until staling occurs. The ideal bread packaging material must be attractive, strong and inexpensive. It must have adequate moisture barrier properties to improve the shelf-life, able to run on automatic machinery and lastly should protect the shape of the product.

Hence the packaging material selected must conserve the moisture content, prevent staling and keep the bread in a fresh condition as long as possible.

The ideal bread packaging material must:

- be attractive
- maintain adequate shelf-life
- run on automatic machinery
- be strong
- be inexpensive
- be an adequate moisture barrier, and
- protect the shape of the product

Packages for Biscuits

Since most of the bakery products are packed on automatic form-fill-machine which run at fairly high speeds, the packaging material selected must be capable of running efficiently on these machines.

Packaging Materials

(a) Biscuits

A wide range of packaging materials is used to pack biscuits. Since paper cartons, tins have lost out to flexible packaging materials as the packaging medium, focus is now on the latter. A variety of flexible packaging materials are used for packing biscuits due to advantages such as functionality, lower cost, printability, light weight, savings in freight and other such factors.

1. **Flexible Packaging Materials**: These are used as wrappers, pre-formed pouches or form-fill pouches. The oldest flexible film to be used was cellophane because of its excellent gas barrier properties and heat sealability. MST, MSAT, Coated Cellophane (MXXT) offer excellent moisture barrier, heat sealability and gloss. Cellophane became less popular when it became too expensive and with the introduction of new materials with better properties. Another material, which is widely used is Biaxially Oriented Polypropylene film commonly known as OPP. For less demanding applications OPP monofilm is used while for higher quality products, duplex OPP or OPP combinations (pearlized or metalized) such as OPP/PE, OPP/PET etc. are used. Today most of the biscuits are packed in flexible laminates of composite structures, where every component fulfills a specific function. These laminates have desirable properties such as moisture barrier, gas barrier heat sealability, printability characteristics, high production and overall economy.

The different types of plastic films and its uses are tabulated below.

Plastic Films	Uses
Low density polyethylene (LDPE)	Performed pouches
Polypropylene (PP)	Preformed pouches
Biaially Oriented Polypropylene (BOPP)	Plain or pearlised films as overwraps preformed pouches, pillow pouches on FPS machines.
Polyester/LDPE laminates	Preformed pouches or FFS pillow packs
Metallised polyester/poly	FFS pillow packs
Paper/Foil/Poly	FFS pillow packs

2. **Thermoformed Plastic Trays**: Thermoformed plastic trays of polyestyrene or PVC with multiple cavities are used to pack assorted biscuits, pastries, cookies etc. They are closed with a snap-on lid or overwrapped or shrink-wrapped or sealed with a lidding material. The products rest nicely in the compartments and make a good presentation. Use of active packaging with oxygen absorbent and antimicrobial properties for bakery products helps to significantly increase the shelf-life and maintain the original quality of the product. PVDC coated nylon, polyester, LDPE, PP, ethylene vinyl alcohol, polystyrene are examples of flexible packaging material used with active sachet.

(b) Bread

Traditionally, bread in India was packed in waxed paper wrappers. The search for lower cost over wrapping materials led to the use of polyethylene film and nearly 80% of all bread is now packed in plastics films such as LDPE, LLDPE-LDPE and PP. Also, auto bagging machines require high slip PE resin i.e. pouches with good openability. LLDPE/ LDPE bags of 1 to 1.5 mm thickness secured by plastic clip or twisted wire ties are normally used.

Packages of Sliced Bread

(c) Cakes, Pastries, Doughnuts

These products are available in various sizes, shapes and forms. Since these products contain high moisture content they are prone to mould growth and hence the packaging material selected should not encourage mould growth. The packaging material used is Polypropylene (PP), Cast Polypropylene (CPP), Poly Vinyl Chloride (PVC) etc., while the choice of the film depends upon the machinability and economics required.

The Packaging Styles

There are several popular wrapping styles, which are applied widely to a variety of biscuits (of all shapes and sizes).Biscuits packed using the following two wrapping styles must be of common size and shape with a certain consistency and rather narrow tolerances in their dimensions. Standard wrapping machines can be used.

Endfold Wrapping

This wrapping style is the classic, traditional biscuit wrapper. A portion of biscuits standing on edge is roll – wrapped or fold wrapped into a heat sealable film. The longitudinal packet seal is sealed tightly in a fin seal style. The packet ends are folded neatly and heat-sealed. Due to the neat and tight surrounding of the film, this packet gives utmost mechanical protection and acceptable barrier properties for hard and semi -hard biscuits and many other cracker types. Enfold wrapping is considered the most effective in terms of presentation by many marketing specialists - not only due to neat and impeccable shape, but also due to its ability to clearly distinguish the product amongst the host of pillow pack items on the retail shelves.

Pillow Pack Wrapping

This is the standard wrapping style for smaller biscuit packs (snack packs/ single serve packs) containing one or more piles of biscuits. In addition, pillow pack wrapping is used for bigger packets with products standing on edge (Slug wrapping) as well. In this configuration, it often serves as a primary wrapper, to be over-wrapped by a carton to improve presentation and acceptance. The main advantage of pillow packs on edge, is its flexibility with regard to the slug length. For instance, it allows the machine to automatically adjust the length during wrapping by means of tendency controlled check weighers. This feature ensures the highest weight accuracy. Additionally, the pillow packs typical fin seal style sealing is somewhat tighter than the enfold wrap. This disadvantage of pillow pack slug wrapping is its limited mechanical product protection due to its rather loose packing. Further, the presentation of products packed using the pillow pack style is considered by most to be less attractive than enfold packets.

Square/Rectangular Biscuits in
Plastic Laminates - Flo Pack

Packing for Odd-sized Biscuits

Besides enfold wrapping and pillow pack wrapping, which by the way cover about 85–90% of all biscuit products, there are some specialty biscuits with their own unique wrapping needs. These include an assortment of small cocktail crackers filled in bags by vertical FFS, machines and cookies of uneven sizes whose tolerance do not allow a standard wrapping. The latter are automatically or manually loaded into decorated trays and subsequently over-wrapped on pillow pack machines.

Bakery products include items of different packaging requirements, which are met by a range of plastic materials in the form of films, laminates and thermoformed trays. These materials provide adequate protection against moisture loss/gain, retain the taste and aroma, and are hygienic and safe for food contact. Other additional properties such as machinability, printability and cost effectiveness make them the ideal choice for a package.

Packaging of Confectionery Products

Confectionery items are commonly consumed by the populace and generally used for taste and desire and becoming very popular especially amongst children and youth. There are various types of confectionery items available in market i.e. cakes, pastries, doughnuts, candies, wafers, chips, chewing gum and chocolates. There are some leading industries in the field of confectionery products are Cadbury, Nestle, Perfetti, Wrigley, Parle and Amul etc. A package intended for sugar and chocolate confectionery has to perform several functions during distribution, storage and sales. Essentially, the package has to preserve the quality attributes of the product and afford protection against chemical and microbiological deteriorative reactions.

For sugar confectionery items and chocolates, the major functional packaging requirements include protection from: (a) Dust, dirt and other contaminating agents (b) Moisture/water vapour pickup or loss resulting in sugar and fat bloom, stickiness, hardening and desiccation. (c) Rancidity due to interaction with moisture and oxygen. (d) Colour and aroma loss and tainting. (e) Physical damages like dusting, breakage and loss of shape. In addition to the above, the packaging material should be amenable to run well on machines, should be hygienic and do not cause any health problem. Currently the major addressable problem is that it should be eco-friendly and easy to use and dispose-off.

Role of Water Activity (aw) in Confectionery Products

The end of shelf life due to moisture loss or gain, with subsequent changes in textural and other properties, is often the main problem in confections. Thus, an understanding of water activity is important for control of shelf life and stability. Water activity in confections generally falls below the critical values for microbial growth, with few exceptions. Water activity is influenced by the presence of dissolved sugars, other sweeteners (e.g., polyols), salts (e.g., caramel), and humectants in confections. Microbial growth is directly related to aw, with certain types of microbes unable to grow when water activity is below some critical value. The following table shows the water activity range of different confectionery items is listed below in Table 14.1.

Table 14.1 Range of water content and water activity (aw) in confections

Category	Crystallinity (%)	Moisture (%)	aw
Hard candy	0–2	2–5	0.25–0.40
Caramel, fudge, toffee	0–30	6–18	0.25–0.40
Chewy candies	0–10	6–10	0.45–0.60
Nought	0–20	5–10	0.40–0.65
Marshmallow	0–20	12–20	0.60–0.75
Gummies and Jellies	0	8–22	0.50–0.75
Jams	00	30–40	0.80–0.85
Fondants and Creams	35–55	10–18	0.65–0.80
Chewing gum	30–40	3–6	0.40–0.65
Soft panned coating	60–75	3–6	0.40–0.65
Hard panned coating	80–95	0–1	0.40–0.75
Tablets and lozenges	75–95	0–1	0.40–0.75

Growth of Microorganisms Dependent on Water Activity (Aw)

Microbial growth also decreases with reduction in water activity. Majority of bacteria and many yeasts stop growing below water activity of 0.88. Only few osmophilic yeasts and molds grow below water activity of 0.7 and no microorganisms can grow below the water activity of 0.6. Growth range of microorganisms in terms of water activity.

Table 14.2 Water activity and growth of microorganism in confectionery products

Water Activity	Microorganisms that can grow	Confections
>0.88	Normal bacteria and pathogens, many yeasts	Ganache, very soft fondant
0.80–0.88	Normal molds and yeasts	Soft fondant, soft jellies
0.70–0.80	Molds and yeasts	Fondant, fudge, jellies, grained noughats, marshmallow
0.60–0.70	Osmophilic yeasts, some molds	Fudge, fondant, hard jellies, noughat, soft caramel
< 0.60	None	Caramel, toffee, jellies, gum, hard candy, chocolate etc.,

Package Forms

Different package forms are used for packaging of chocolate and other confectionery items. The different types of package forms are: Consumer unit packages, Shipping/bulk containers, Flexible packages, Semi-rigid packages, Rigid containers, Conventional bags, Corrugated fibreboard boxes, sacks and boxes.

Consumer Unit Packages

Flexible packages

The basic styles of flexible pouch/bag used to contain sugar confections and chocolates are - flat and pillow type, satchel bottom and stand-up types. The materials of construction comprise functional papers, plastics, aluminium foils, metalized films and composites of these materials. Although conventionally, different kinds of papers such as Poster, grease-proof and glassine, vegetable parchment and even sometimes news-paper are used for making bags, newer materials include polyolefins, polyethylene terephthalate (PET or polyester), polyamide (Nylon), aluminium metalized plastic films and co-extruded structures are being increasingly used.

Since polyester, polyamide and metallised films (PET, PP and PA) films possess very good barrier properties towards oxygen and aromas, but unsupported, do not provide good heat-seals. They are generally combined with polyethylene and its modifications to provide sealability and inertness. A recent development in the flexible packaging field is the availability of co-extruded structures. A most common combination suitable for sugar confectioneries is an outer polyethylene - Low-density or linear low density (LDPE or LLDPE), core or middle layer of polyamide (Nylon-6) or ethylene-vinyl alcohol (EVOH) copolymer and inner or contact layer of Ionomer or Ethylene-Acrylic Acid (EAA) copolymer film.

To enhance the shelf life of the products, either vacuum or gas packaging can be resorted to. The former technique is suitable for packing sweets having rigid structure such as sohan papdi and Mysore pak, while gas packaging is better suited for softer confections. For these applications, materials such as plain and metalized PET with PE, or copolymer of HD-LDPE, Nylon and EVOH based co-extruded films and polyvinylidine chloride (PVDC) copolymer, coated PP would be the better choice.

Semi-rigid containers

These comprise folding cartons, set-up boxes, lined-folding cartons and thermoformed containers.

Collapsible folding cartons of tray-type with coated or laminated paperboard base are extensively used to package dairy food-based sugar confections. These cartons with outer embellishments are best suited for gift and display applications. The liner material may be PE-wax-EVA blends or PVC or PET films. Set-up boxes of either half or full telescopic type having a inner glassine liner are economical and provide good physical and mechanical protection.

Lined folding cartons system is of bag-in-box type where an inner pouch is lined (fixed) to the outer paperboard carton. The selection of the material of the pouch is decided by the functions required, economics and marketing requirements. Materials such as paper/PE, *PET/PE,* paper/ Al-foil/PE and the almost ultimate choice, *PET/*Al foil/PE are used. Provisions for reclosing, reduction of headspace volume and such features can be incorporated.

Thermoformed containers include blister packs, single and multi-cavity trays, thin-walled containers with lids etc. These are produced by the process of thermoforming by vacuum, pressure or matched techniques. For packing sweets, thermoformed tray type containers are better suited. For multi-coating trays, the number, shape and size of cavities is determined by the product to be packed. Such trays are useful when a number of similar or assorted items are packed.

Eco-friendly packages

Bio-containers or eco-friendly packages based on natural materials such as leaves of banana can also be used to contain and distribute sugar-based confectioneries. The processes developed by CFTRI to manufacture these involve only heat-treatment without recourse to any additional adhesive or chemical treatments.

Rigid packaging systems

Among the metal containers, the conventional tinplate cans are being used to process *rasagolla* and *gulab jamun* in syrup. Tinplate cans are available in various standard sizes. For flat sweets, 100 g cans are preferred. For gas flushing applications, formed cans with aluminium top cans are used. Newer metal containers include differentially coated cans, chromium coated (Tin-free-steel) cans. The provision of ring-pull ends (Easy-open-end; EOE) facilitates easy opening for consumers. Aluminium containers are made by different techniques. These are available in circular, oval, rectangular or any fancy shapes and can be decorated in an unlimited range of design and colour variations. EOE ends with reclosable polyethylene lids are finding greater applications for sugar and dairy products.

Composite containers

These are made of paperboard body and metal or plastic ends. The container body may either be spirally or convolutedly wound with fibre board lined with aluminium foil. Composite can having a body material of 25 mm PE/Paper-board/0.009 mm aluminium foil/37 mm LDPE are well suited to package sweet meats.

Shipping containers

Corrugated fibreboard boxes (CFB) are being employed as exterior containers for packing unit packs both for inland and export markets. They can be used up to a maximum weight of contents of 75 kg. The BIS specifies, depending on the maximum of mass of contents:

(a) Maximum combined internal dimensions (LxWxD),

(b) Minimum bursting strength of the boards

(c) Combined liner grammage

(d) Moisture absorptiveness value (Cobb value).

The style of the box is decided by the contents, protection required and marketing destinations.

Packaging and Storage Studies

For designing packages, the primary requisite is the knowledge of relationship between moisture content and equilibrium relative humidity (ERH) (and hence water activity) denoted through moisture sorption isotherm.

Sohan papdi

This confection normally has a shelf-life of about 12 days with critical moisture content of 3% corresponding to 0.3 water activity. Chemical and sensory analysis carried out on BHA-treated packed product has indicated that the product could be stored well for 20 days in PET/PE, 120 days in HDPE and 225 days in paper/foil/PE pouches and rigid metal container.

Sohan halwa

This product with permissible moisture pick up of 1.7 percent and stored at normal environmental conditions has revealed storage life of 30 days in LDPE, 120 days in HDPE and 180 days in foil-laminate pouches. The shelf life has also been determined in package of tagger-top tinplate containers.

Milk peda

Studies on buffalo milk peda with preservative at two concentrations were carried out. Sample containing 0.002% sorbic acid extended the storage up to 9 days at 30°C/70% RH and 37 days at 7°C/*90%* RH conditions. Sorbic acid added at 0.05% was effective against chemical and microbiological deterioration up to 50 days at 7°C. Milk *peda,* having 0.84 water activity or 14.0% moisture packed in laminates pouches of PET/PE and PET/Al foil/PE and co-extruded LLDPE/PA/EAA has indicated that ambient-air packaging did not extend the shelf-life. Usage of free-oxygen absorber sachets extended the life up to 42 days.

Extensive studies have been carried out on hard-boiled sweets, toffees and chocolates on moisture sorption characteristics. Packaging and storage studies have revealed that sucrose in amorphous state in hard boiling have critical moisture content in the region of 1.2–1.5%. Modified and plain toffees had higher critical values, which was in the region of 4–6%. Plain and milk chocolates having moisture contents of 0.7% and 1.33% respectively, equilibrating to 0.64 water activity was found to be critical with respect to maintenance of good colour, aroma and texture.

References

Kumar K.R. Packaging Aspects of Confectionery Items. Indian Food Industry, Jan-Feb '92, Vol. 11(1).

Ravichandran, P. and Kumar K.R. Relationship Between Moisture Sorption and Texture Characteristics of Sugar Confectionery Indian Food Packer, May–Jun 2001

www. indiainfoline.com/comp/cadb/1401.html

15

Quality Assessment and Standard Specifications of Confectionery Products

The art of confectionery manufacture date back to over 3000 years as per Egyptian records. Traditionally honey, boiled and concentrated sugar cane juices have been used in confectioneries. In modern times, alternative sweeteners from sources like corn and starches constitute an integral part of confectionery manufacture.

Owing to the huge variation in finished products as well as raw materials in food sector, Quality control has an extremely important role to play. Quality control aims at controlling variation to within a tolerable level by taking corrective actions. Statistical and non-statistical techniques are employed to measure, analyze and control variation in food products. Statistical process control, acceptance sampling and visual inspection are widely used in food and allied sectors.

Maintenance of quality and manufacture of a standard confectionery should involve a very close control of raw materials, instrumentation, and statistical analysis of finished product. Statistical evaluation with respect to finished product includes weight control, sensory evaluation and packaging tests. The quality standards are set by the quality control departments. Chemists may undertake preliminary analysis through visual inspection. Visual inspection is usually combined with sampling. This gives an idea of variation, cleanliness, infestation etc. in the sample.

Confectionery Products

Confectionery products cover a wide spectrum of products ranging from candies, chocolates, cakes, and other sweet foods

Classification

Confectioneries may be broadly classified as:

- Sugar confectionery-These are not covered with chocolate. Examples include hard candy, toffees, fudge, fondants, jellies, pastilles etc.

- Chocolate confectionery-these are usually sugar confectioneries covered with chocolate and include chocolate bars and blocks.
- Flour Confectionery-baked fancy cakes that are either ice or chocolate covered
- Traditional Indian sweetmeats

Confectionery Ingredients

1. **Flours:** High flour temperatures can adversely affect the product quality. Therefore proper control of temperature must be exercised during steps like milling and pneumatic handling.
2. **Sugars, Glucose syrups and other Sweeteners:** Sugar (sucrose, saccharose, fructose, dextrose) is the main ingredient of confectionery products. Sugars used for commercial purpose may be either of cane or beet origin. Sugars may be used in various forms including refined sugars, various grades of brown sugar, syrups and invert sugar and others. Sugar alcohols like sorbitol and mannitol and non nutritive sweeteners including saccharin, cyclamates, acesulfame K, aspartame etc. have also been used in confectionery manufacture. Sugar must be dry and should meet the relevant particle size requirements and color specifications.
3. **Confectionery Fats:** Raw fat obtained from various sources is subjected to refining prior to use in food. Refining includes neutralization, bleaching and deodorization. Commercial edible oils are obtained from coconut, palm, peanut, soybean, rapeseed, cottonseed, sesame seed, sunflower, olive, corn etc. Properties of fats that are of importance are consistency, aeration, resistance to oxidation, melting temperature. Physical properties of the fats muct be within the acceptable range as per relevant regulation
4. **Milk and milk products:** Milk and milk products provide the desirable flavor to the confectionery products. Lactose, butter, condensed milk, evaporated milk, milk powder, whey powders, liquid milk are the various forms in which milk is generally used in confectionery. Other forms are malted milk and cultured milks
5. **Egg albumen and other aerating agents:** All soluble proteins show aeration property .Apart from albumen other aeration agents that are used include casein, whey, soy protein, skimmed milk powder, and gelatin. Quality of egg products must be evaluated from both bacteriological and functional property viewpoints.
6. Gelatinizing agents, gums, waxes-agar-agar, alginates, carrageenan, gum Arabic etc.

7. Starches, soy flour, soy protein; Natural starches: corn, wheat, potato, arrowroot, tapioca, modified starches
8 Fruits, preserved fruits, jams, dried fruits. Fruits should be wholesome, of uniform size and confirm to cleanliness standards.
9. Nuts-almonds, cashew nuts, coconut, brazil nuts, chest nuts

Quality Control

As per ISO 9000:2000, quality has been defined as "The totality of features and characteristics of a product, process or service that bear on its ability to satisfy stated or implied needs".

FAO defines Quality control as a planned system of activities whose purpose is to provide a quality product.

Quality control is also known as quality inspection and thus as suggestive of the name it involves checking for quality standards at various steps of manufacturing.

Quality control is carried out in the following three areas of operation:

1. Raw ingredients.
2. Process of manufacture.
3. Inspection of finished product.

Control of Raw Materials

Preliminary testing is done to determine acceptability of delivery. Acceptance or rejection of the consignment is generally decided by the analytical chemists. This includes visual inspection. Raw material that fails to meet the prescribed standards is rejected. Sampling should be a representative of the bulk. On getting due approval for further processing from the concerned personnel, the containers are opened in batch rooms or stores and weighed amounts are sent for production process.

Type of Raw Material

- If the source of supply and manufacture is known and reliable, a superficial examination is sufficient.
- All packages, drums or container should be marked appropriately to avoid confusion-glucose syrups (degree of conversion), starches(crème depositing or thin boiling) an fats (melting point)
- Essential oils, flavors, spices, nuts, dried fruits, egg albumen are checked for quality differently.
- Essential oils are subjected to flavor tests, refractive index, specific gravity, purity tests

- Spices are tested through flavor tests, microbiological tests, extraneous filth
- In case of Cocoa beans, "cut" test is performed to check if they are properly fermented
- Nuts and dried fruits are analyzed for flavor, foreign matter, moisture content
- Sugars when used in the form of syrups should be checked for parameters like pH, total solids and invert sugar contents, color, temperature

The Supplier

- More known is the supplier less is the degree of inspection that needs to be carried out.
- Issues like specifications, keeping periods, type of packing, methods of testing should be discussed with the supplier.
- The supplier should have a quality control system for the manufacture.

Reciept and Preparation for Production

- Sugar, glucose syrup, mixed syrups, fats are usually delivered in tank truck and rail tank cars. Production personnel are also responsible for quality control.
- A rapid inspection of a representative sample drawn from the tanker s advised at the arrival of the tanker
- Some physical tests like specific gravity or moisture are carried out at such stage.
- Sample may be tasted and examined visually for foreign matter.
- Detection of defective or material of unacceptable quality must be reported by the personnel handling the opening of containers.
- Proper testing of alternative raw materials should be done when introduced.
- When using a substitute material, its purity must be checked prior to incorporation into the production process. Initial visual inspection servers as a quick means of inspection

Process Control

- The process of manufacture can be either a batch type production processes or continuous type production lines.
- Batch processes are handled manually .Therefore quality greatly depends on the efficiency and reliability of the workers
- Larger the product quantity greater are the number of batches and thus more likely are the chances of inter batch variation.

- It is not practically possible to do sufficient checking of each batch. Usually one analytical testing is done for each batch like moisture content.
- Data to formulate continuous processes are obtained from automated bath processes.
- Time of cooking and cooling are longer in a batch processes compared to continuous processes.

Type of process used depends on the type of end product:

1. Fudge, caramel: Maillard reaction between the milk proteins and sugar components leads to desirable flavor and color generation. Short duration continuous processes require substitution of additional caramelizers to complete the process
2. Pectin jellies: Appreciable sugar inversion is attainable in batch boiling. Sugar inversion is required to generate acidity which is a desirable in case of pectin jellies.
3. Microbial contamination: ferments, molds and enzymes are present in raw materials like cocoa beans, nuts, dried fruits, egg albumen. These microbes are not inhibited in batch processes unlike when added towards the end of continuous process.

Finished Product Inspection

- Adoption of the in- line control eliminates the need for frequent inspection.
- Generally the finished product is analyzed for three significant parameters namely: Appearance, taste and weight.

1. Appearance

Appearance involves both the product appearance and the packaging. The finished product should have satisfactory appearance. A "pattern" consisting of the packed box of the product being manufactured is provided to the inspection staff. The pattern is approved by the marketing and quality control department.

2. Taste Checks

Samples taken for weight control are also subjected to sensory evaluation. Various sensory evaluation methods may be employed for this like the duo trio test, triangle test, hedonic scale rating etc.

Sensory tests are classified as follows:

Discrimination: They have the objective of determining if differences exist between two or more products differences exist between two or more products. Type of questions that may be asked in discrimination sensory tests are "Is product A identical to product B?", " Is product A identical to product

B?", "Find the two similar products among the three samples provided.", "Find the odd sample among the three samples Find the odd sample among the three samples provided"

Description
These have the objective of describing characteristics of a product and/or measuring any characteristics of a product and/or measuring any differences that are found between products .It answers questions like, "What does this product taste like? What does this product taste like?", "What are the three most important texture attributes you perceive in this product? ", "For which sensory attributes are the differences For which sensory attributes are the differences between product A and B most marked?

Preference or Hedonics
These tests describe liking or acceptability of a product. Questions may be of the type "Do you like this product? ", "How much do you like this product on a scale of 1 to 10, where 1 = dislike product on a scale of 1 to 10, where 1 = dislike extremely, and 10 = like extremely? ","Is product A better than product B?"

Weight Control

- Overweight pieces reduces profit
- Underweight products liable for legal offence
- Frequent checking is essential
- Weight charts are prepared
- A weight distribution chart helps to know the points where control is needed

Finished Packs

Package testing is an important aspect of quality control since the keeping quality of finished product depends greatly on the quality of the packaging material. The following tests are recommended for package testing:

1. Flexible packaging- yield, dimensions, coefficient of friction, seal strength wherever applicable
2. Foil laminates-yield, sealing temperature
3. Waxed paper-yield, surface wax, total wax
4. Cartons, tins, etc-
5. Others-weight of a unit wrapping, goodness of print, odor and taint barrier property.

Shelf-Life Study

Rate of deterioration depends on combination of factors like temp of storage, ingredients quality, recipe used, nature of packaging. Shelf life refers to the time period for which the product retains its original quality at relevant storage conditions. Storage beyond this time period, the product becomes unsaleable. To determine the keeping life accelerated keeping tests are done in place of prolonged storage tests to save time.

- Prolonged storage tests-carried out under average shop conditions.
- Accelerated keeping tests involve subjecting the samples to different storage conditions along with different packaging. The samples are then tasted at appropriate intervals. Thermostatically controlled incubators at 18°C, 23°C, 27 °C, 29.5 °C are preferred. These are carried out for one to two months under special storage conditions

  Special storage conditions:
  Tropical conditions - 29 °C, 85–90% RH
  Cool storage - 7 °C, 10°C
  Cold storage - 7 °C
- Destructive testing: Sample is subjected to conditions of fluctuating temperatures. The maximum and minimum conditions used imitate the most severe tropical conditions.

Microbial Testing

Raw materials

Raw materials should be free from pathogens. This necessitates microbial testing of the raw materials on reception. The compulsory microbial standards for raw materials are *Salmonella*, negative, E/coli, negative, *Enterobacteriaceae*, negative in 1g, total plate count-less than 5000, molds and yeasts-max 50 per gram, negative lypolytic activity in cocoa , egg albumen-negative.

Process Control

Risk of contamination is most associated with intermediate steps like soaking of gelatin .

Freshly prepared solutions must be used. Equipments used for such solutions must be washed and sterilized.

Finished Products

Proper control of raw material and processing minimizes risk of contamination in finished product. Handling during packaging is a hazard

Complaints

Complaints from consumers must be addressed properly. Indications of microbial contamination are rancidity, off flavor generation, microbial liquefaction ets. Relevant samples must be analyzed for the root cause.

Chemical Analysis

1. Laboratory Practice
2. Sugar analysis
3. Moisture Analysis
4. Protein Analysis
5. Fat Analysis-Traditional method: Soxhlet apparatus. However accurate results can't be obtained by this method as some amount of bound fat remains unextracted and needs an additional step of acid hydrolysis. A modified and more rapid extraction system which is used now is the Fosslet fat analyzer. HCl is added following a reaction between perchloroethylene to facilitate acid hydrolysis. The extraction is completed by finally adding plaster of paris to absorb the aquous phase. This method gives very accurate results in the analysis of toffees and caramels.
6. Aflatoxins-Quantitative results are obtained by analyzing the sample for aflatoxins using HPLC with a fluorescence detector. Monoclonal antibodies based test kits have also been brought to use for detection of aflatoxins. Aflatoxins are preferentially absorbed onto mini columns containing specific antibodies (e.g. Aflatest 10). Aflatoxins are eluted and absorbed onto fluorosil tips and quantified against standard in UV viewing cabinet.
7. Viscosity-U-tube, falling ball, cup or torsional viscometers have been used traditionally for viscosity measurement. However, digital rotational viscometer is the most widely used instrument for measuring viscosity. The instrument gives instant results on receiving temperature input. The results are expressed in poise.
8. Particle Size-Retention on standard sieves or by microscopy. Other methods are tedious and expensive
9. Acid Content-Volumetric method where the diluted sample is titrated against sodium hydroxide. Phenophthalein indicator is used to determine end point. Calculations are done using the relevant correction factor.

Modern Methods

Some modern methods that can be employed for analysis of confectionery products:

1. Nuclear magnetic resonance (NMR)

Hydrogen nuclei absorb radiofrequency energy in presence of magnetic field. NMR spectroscopy can be used to measure water content, fat content of dry products like chocolates. It also gives the solid: liquid ratio in products at varying temperature .This helps to predict the performance of fats.

2. Near Infrared (NIR)

NIR technique helps in very rapid determination of contents of fat, protein and moisture in the sample. This technique has disadvantage that for every analysis hundred different standards need to be calibrated.

Standard For Chocolate and Chocolate Products (Codex Stan 87-1981, Rev. 1 - 2003)

Scope

1. The standard applies to chocolate and chocolate products intended for human consumption and listed in Section 2. Chocolate and chocolate products shall be prepared from cocoa and cocoa materials with sugars and may contain sweeteners, milk products, flavouring substances and other food ingredients.

2. Description and Essential Composition Factors

Chocolate is the generic name for the homogenous products complying with the descriptions below and summarized in Table 1. It is obtained by an adequate manufacturing process from cocoa materials which may be combined with milk products, sugars and/or sweeteners, and other additives listed in section 3 of the present standard. Other edible foodstuffs, excluding added flour and starch (except for products in sections 2.1.1.1 and 2.1.2.1 of this Standard) and animal fats other than milk fat, may be added to form various chocolate products. These combined additions shall be limited to 40% of the total weight of the finished product, subject to the labeling provisions under Section 5. The addition of vegetable fats other than cocoa butter shall not exceed 5% of the finished product, after deduction of the total weight of any other added edible foodstuffs, without reducing the minimum contents of cocoa materials. Where required by the authorities having jurisdiction, the nature of the vegetable fats permitted for this purpose may be prescribed in applicable legislation.

2.1 Chocolate Types (Composition)

2.1.1 Chocolate Chocolate (in some regions also named bittersweet chocolate, semi-sweet chocolate, dark chocolate or "chocolat fondant") shall contain, on a dry matter basis, not less than 35% total cocoa solids, of which not less than 18% shall be cocoa butter and not less than 14% fat-free cocoa solids. 2.1.1.1 Chocolate a la taza is the product described under Section 2.1.1 of this Standard and containing a maximum of 8% m/m flour and/or starch from wheat, maize or rice.

2.1.2 Sweet Chocolate Sweet Chocolate shall contain, on a dry matter basis, not less than 30% total cocoa solids, of which at least 18% shall be cocoa butter and at least 12% fat-free cocoa solids.

2.1.2.1 Chocolate familiar a la taza is the product described under Section 2.1.2 of this Standard and containing a maximum of 18% m/m flour and/or starch from wheat, maize or rice.

2.1.3 Couverture Chocolate Couverture Chocolate shall contain, on a dry matter basis, not less than 35% total cocoa solids of which not less than 31% shall be cocoa butter and not less than 2.5% of fat-free cocoa solids.

2.1.4 Milk Chocolate Milk Chocolate shall contain, on a dry matter basis, not less than 25% cocoa solids (including a minimum of 2.5% fat-free cocoa solids) and a specified minimum of milk solids between 12% and 14% (including a minimum of milk fat between 2.5% and 3.5%). The minimum content for milk solids and milk fat shall be applied by the authority having jurisdiction in accordance with applicable legislation. "Milk solids" refers to the addition of milk ingredients in their natural proportions, except that milk fat may be added, or removed. Where required by the competent authority, a minimum content of cocoa butter plus milk fat may also be set.

2.1.5 Family Milk Chocolate Family Milk Chocolate shall contain, on a dry matter basis, not less than 20% cocoa solids (including a minimum of 2.5% fat-free cocoa solids) and not less than 20% milk solids (including a minimum of 5% milk fat). "Milk solids" refers to the addition of milk ingredients in their natural proportions, except that milk fat may be added, or removed. Where required by the competent authority, a minimum content of cocoa butter plus milk fat may also be set.

2.1.6 Milk Chocolate Couverture Milk Chocolate Couverture shall contain, on a dry matter basis, not less than 25% cocoa solids (including a minimum of 2.5% non-fat cocoa solids) and not less than 14% milk solids (including a minimum of 3.5% milk fat) and a total fat of not less than 31%. "Milk solids" refers to the addition of milk ingredients in their natural proportions, except that milk fat may be added, or removed.

2.1.7 Other chocolate products

2.1.7.1 White Chocolate White Chocolate shall contain, on a dry matter basis, not less than 20% cocoa butter and not less than 14% milk solids (including a minimum milk fat in a range of 2.5–3.5% as applied by the authority having jurisdiction in accordance with applicable legislation). "Milk solids" refers to the addition of milk ingredients in their natural proportions, except that milk fat may be added, or removed. Where required by the competent authority, a minimum content of cocoa butter plus milk fat may also be set.

2.1.7.2 Gianduja Chocolate

"Gianduja" (or one of the derivatives of the word "Gianduja") Chocolate is the product obtained, firstly, from chocolate having a minimum total dry cocoa solids content of 32%, including a minimum dry non-fat cocoa solids content of 8%, and, secondly, from finely ground hazel nuts such that the product contains not less than 20 % and not more than 40% of hazel nuts. The following may be added: - (a) milk and/or dry milk solids obtained by evaporation, in such proportion that the finished product does not contain more than 5% dry milk solids; - (b) almonds, hazelnuts and other nut varieties, either whole or broken, in such quantities that, together with the ground hazel nuts, they do not exceed 60% of the total weight of the product.

2.1.7.3 Gianduja Milk Chocolate

"Gianduja" (or one of the derivatives of the word "Gianduja") Milk Chocolate is the product obtained, firstly, from milk chocolate having a minimum dry milk solids content of 10% and, secondly, from finely ground hazelnuts such that the product contains not less than 15 % and not more than 40% of hazelnuts. "Milk solids" refers to the addition of milk ingredients in their natural proportions, except that milk fat may be added or removed. The following may be added: Almonds, hazelnuts and other nut varieties, either whole or broken, in such quantities that, together with the ground hazelnuts, they do not exceed 60% of the total weight of the product. Where required by the competent authority, a minimum content of cocoa butter plus milk fat may also be set.

2.1.7.4 Chocolate para mesa

Chocolate para mesa is unrefined chocolate in which the grain size of sugars is larger than 70 microns.

2.1.7.4.1 Chocolate para mesa

Chocolate para mesa shall contain, on a dry matter basis, not less than 20% total cocoa solids (including a minimum of 11% cocoa butter and a minimum of 9% fat-free cocoa solids)

2.1.7.4.2 Semi-bitter chocolate para mesa

Semi-bitter Chocolate para mesa shall contain, on a dry matter basis, not less than 30% total cocoa solids (including a minimum of 15% cocoa butter and a minimum of 14% fat-free cocoa solids). 2.1.7.4.3 Bitter chocolate para mesa Bitter Chocolate para mesa shall contain, on a dry matter basis, not less than 40% total cocoa solid (including a minimum of 22% cocoa butter and a minimum of 18% fat-free cocoa solids).

2.2 Chocolat Types (Forms)

2.2.1 Chocolate Vermicelli and Chocolate Flakes

Chocolate Vermicelli and Chocolate Flakes are cocoa products obtained by a mixing, extrusion and hardening technique which gives unique, crisp textural properties to the products. Vermicelli is presented in the form of short, cylindrical grains and flakes in the form of small flat pieces.

2.2.1.1 Chocolate Vermicelli / Chocolate Flakes

Chocolate Vermicelli / Chocolate Flakes shall contain, on a dry matter basis, not less than 32% total cocoa solids, of which at least 12% shall be cocoa butter and 14% fat-free cocoa solids.

2.2.1.2 Milk Chocolate Vermicelli / Milk Chocolate Flakes

Milk Chocolate Vermicelli / Milk Chocolate Flakes shall contain, on a dry matter basis, not less than 20% cocoa solids (including a minimum of 2.5% fat-free cocoa solids) and not less than 12% milk solids (including a minimum of 3% milk fat). “Milk solids” refers to the addition of milk ingredients in their natural proportions, except that milk fat may be added, or removed. Where required by the competent authority, a minimum content of cocoa butter plus milk fat may also be set.

2.2.2 Filled Chocolate

Filled Chocolate is a product covered by a coating of one or more of the Chocolates defined in Section 2.1, with exception of chocolate a la taza, chocolate familiar a la taza and products defined in section 2.1.7.4 (chocolate para mesa), the centre of which is clearly distinct, through its composition, from the external coating. Filled Chocolate does not include Flour Confectionery, Pastry, Biscuit or Ice Cream products. The chocolate part of the coating must make up at least 25% of the total weight of the product concerned. If the centre part of the product is made up of a component or components for which a separate Codex Standard exists, the component(s) must comply with the applicable standard.

2.2.3 A Chocolate or Praline

A Chocolate or Praline designates the product in a single mouthful size, where the amount of the chocolate component shall not be less than 25% of the total weight of the product. The product shall consist of either filled chocolate or a single or combination of the chocolates as defined under Section 2.1, with exception of chocolate a la taza, chocolate familiar a la taza and products defined in section 2.1.7.4 (chocolate para mesa).

Table 1 Summary Table of Compositional Requirements of Section 2[1]

(% calculated on the dry matter in the product and after deduction of the weight of the other edible food stuffs authorized under Section 2)

Products	**Constituents (%)**						
2. Chocolate Types	**Cocoa Butter**	**Fat-free Cocoa Solids**	**Total Cocoa Solids**	**Milk Fat**	**Total Milk Solids**	**Starch/ Flour**	**Hazelnuts**
2.1 Chocolate Types (Composition)							
2.1.1 Chocolate	≥ 18	≥ 14	≥ 35				
2.1.1.1 Chocolate a la taza	≥ 18	≥ 14	≥ 35			< 8	
2.1.2 Sweet Chocolate	≥ 18	≥ 22	≥ 30				
2.1.2.1 Chocolate familiar a la taza	≥ 18	≥ 12	≥ 30			< 18	
2.1.3 Couverture Chocolate	≥ 31	≥ 2.5	≥ 35				
2.1.4 Milk Chocolate		≥ 2.5	≥ 25	≥ 2.5–3.5	≥ 12–14		
2.1.5 Family Milk Chocolate		≥ 2.5	≥ 20	≥ 5	≥ 20		
2.1.6 Milk Chocolate couverture		≥ 2.5	≥ 25	≥ 3.5	≥ 14		
2.1.7 Other Chocolate Products							
2.1.7.1 White Chocolate	≥ 20			≥ 2.5–3.5	≥ 14		
2.1.7.2 Gianduja Chocolate		≥ 8	≥ 32				≥ 20 et ≤ 40

Products	Constituents (%)						
2. Chocolate Types	Cocoa Butter	Fat-free Cocoa Solids	Total Cocoa Solids	Milk Fat	Total Milk Solids	Starch/ Flour	Hazelnuts
2.1.7.3 Gianduja Milk Chocolate		≥ 2.5	≥ 25	≥ 2.5–3.5	≥ 10		≥ 15 et –40
2.1.7.4 Chocolate para mesa							
2.1.7.4.1 Chocolate para mesa	≥ 11	≥ 9	≥ 20				
2.1.7.4.2 Semi-bitter Chocolate para mesa	≥ 15	≥ 14	≥ 30				
2.1.7.4.3 Bitter Chocolate para mesa	≥ 22	≥ 18	≥ 40				
2.2.Chocolate Type s (forms)							
2.2.1 Chocolate Vermicelli/Chocolate Flakes							
2.2.1.1 Chocolate Vermicelli/ Chocolate Flakes	≥ 12	≥ 14	≥ 32				
2.2.1.2 Milk Chocolate Vermicelli/ Milk Chocolate Flakes		≥ 2.5	≥ 20	≥ 3	≥ 12		
2.2.2 Filled Chocolate (See section 2.2.2)							
2.2.3. A Chocolate or Praline (See section 2.2.3)							

[1] "Milk solids" refers to the addition of milk ingredients in their natural proportions except that milk fat may be added or removed.

3 Food Additives

The food additives listed below may be used and only within the limits specified.

Other additives from the General Standard for Food Additives (GSFA) approved list may be used, subject to the authority having jurisdiction in accordance with applicable legislation.

3.1 Alkalizing and neutralizing agents carried over as a result of processing cocoa materials in proportion to the maximum quantity as provided for.

3.2 Acidity Regulators

		Maximum Level
503(i)	Ammonium carbonate	Limited by GMP
527	Ammonium hydroxide	
503 (ii)	Ammonium hydrogen carbonate	
170 (i)	Calcium carbonate	
330	Citric acid	
504 (i)	Magnesium carbonate	
528	Magnesium hydroxide	
530	Magnesium oxide	
501(i)	Potassium carbonate	
525	Potassium hydroxide	
501(ii)	Potassium hydrogen carbonate	
500 (i)	Sodium carbonate	
524	Sodium hydroxide	
500 (ii)	Sodium hydrogen carbonate	
526	Calcium hydroxide	
338	Orthophosphoric acid	2.5 g/kg expressed a P_2O_5 in finished cocoa and chocolate products
334	L-Tartaric acid	5 g/kg in finished products cocoa and chocolate products

3.3 Emulsifiers

		Maximum Level		Products
471	Mono-and di-glycerides of fatty acids			Products described under 2.1 and 2.2
322	Lecithins	GMP		—
422	Glycerol			—
442	Ammonium salts of phosphatidic acids	10 g/kg		—

		Maximum Level		Products
476	Polyglycerol esters interesterified recinoleic acid	5 g/kg	15 g/kg	—
491	Sorbitan monostearate	10 g/kg	in combination	—
492	Sorbitan tristearate	10 g/kg		—
435	Polyoxyethylene (20) sorbitan monostearate	10 g/kg		—

3.4 Flavour Agents

		Maximum Level	Products
3.4.1	Natural flavours as defined in the Codex Alimentarius, and their synthetic equivalents, except those which would imitate natural chocolate or milk flavour[2]	GMP	Products described under 2.1 and 2.2
3.4.2	Vanillin	1 g/kg	Products described under 2.1 and 2.2
3.4.3	Ethyl-vanillin	in combination	Products described under 2.1 and 2.2

[2] Confirme provisoirement.

3.5 Sweeteners

		Maximum Level	Products
950	Acesulfame K	500 mg/kg	Products described under 2.1 and 2.2
951	Aspartame	2000 mg/kg	—
952	Cyclamic acid and its Na and Ca salts	500 mg/kg	—
957	Thaumatin	GMP	—
420	Sorbitol		—
421	Manitol		—
953	Isomalt		—
965	Malitol		—
966	Lactitol		—
967	Xylitol		—

3.6 Glazing Agents

		Maximum Level	Products
414	Gum Arabic (Acacia gum)	GMP	Products described under 2.1 and 2.2
440	Peetin		—
901	Beewax, white and yellow		—
902	Candelilla wax		—
904	Shellac		—

3.7 Antioxidants

		Maximum Level	Products
304	Gum Arabic (Acacia gum)	200 mg/kg	Products described under 2.1 .7.1 Calculated on a fat content basis
319	Tertiary butylhydroquine	200 mg/kg singly or in combination	—
320	Butylated hydroxyanisole		—
321	Butylated hydroxytoluene		—
310	Propylgallate		—
307	α-Tocopherol	750 mg/kg	

3.8 Colours (For Decoration Purpose Only)

		Maximum Level	Products
175	Gold	GMP	Products described under 2.1 .7.1
174	Silver	GMP	—
320	Butylated hydroxyanisole		—
321	Butylated hydroxytoluene		—
310	Propylgallate		—

3.9 Bulking Agents

		Maximum Level	Products
1200	Polydextrose A et N	GMP	Products described under 2.1 .7.1

3.10 Processing Aid

	Maximum Level	Products
Hexane (62 °C – 82 °C)	1 mg/kg	Calculated on a fat content basis

4 Hygiene

4.1 It is recommended that the products covered by the provisions of this standard be prepared and handled in accordance with the appropriate sections of the Recommended International Code of Practice – General Principles of Food Hygiene (CAC/RCP 1-1969, Rev 3-1997), and other relevant Codex texts such as Codes of Hygienic Practice and Codes of Practice.

4.2 The products should comply with any microbiological criteria established in accordance with the Principles for the Establishment and Application of Microbiological Criteria for Foods (CAC/GL 21-1997).

5 Labelling

In addition to the requirements of the Codex General Standard for the Labelling of Prepackaged Foods (CODEX STAN 1-1985 Rev. 1-1991), the following declarations shall be made:

5.1 Name of the Food

5.1.1 Products described under Sections 2.1 and 2.2 of this Standard and complying with the appropriate requirements of the relevant section shall be designated according to the name listed in Section 2 under subsequent section and subject to the provisions under Section 5 of this Standard. The products defined in section 2.1.1 may be described as "*Bittersweet chocolate*", "*Semi-sweet chocolate*", "*Dark chocolate*" or "*Chocolat fondant*".

5.1.1.1 When sugars are fully or partly replaced by sweeteners, an appropriate declaration should be included in proximity of the sales designation of the chocolate, mentioning the presence of sweeteners. *Example*: "X Chocolate with sweeteners".

5.1.1.2 The use of vegetable fats in addition to Cocoa butter in accordance with the provisions of Section 2 shall be indicated on the label in association with the name and/or the representation of the product. The authorities having jurisdiction may prescribe the specific manner in which this declaration shall be made.

5.1.2 Filled Chocolate

5.1.2.1 Products described under Section 2.2.2. shall be designated "Filled Chocolate", "X Filled Chocolate", "Chocolate with X Filling" or "Chocolate with X Centre", where "X" is descriptive of the nature of the filling.

5.1.2.2 The type of chocolate used in the external coating may be specified, whereby the designations used shall be the same as stated under Section 5.1.1 of this Standard.

5.1.2.3 An appropriate statement shall inform the consumer about the nature of the centre.

5.1.3 A Chocolate or Praline

Products in a single mouthful size described under Section 2.2.3 of this Standard shall be designated "*A Chocolate*" or "*Praline*".

5.1.4 Assorted Chocolates

Where the products described under Section 2.1 or 2.2 with exception of *chocolate a la taza*, *chocolate familiar a la taza* and *chocolate para mesa* are sold in assortments, the product name may be replaced by the words "*Assorted*

Chocolates" or "*Assorted filled Chocolates*", "*Assorted Chocolate Vermicelli*", etc. In that case, there shall be a single list of ingredients for all the products in the assortment or alternatively lists of ingredients by products.

5.1.5 Other Information Required

5.1.5.1 Any characterizing flavour, other than chocolate flavour shall be in the designation of the product.

5.1.5.2 Ingredients, which are especially aromatic and characterize the product shall form part of the name of the product (e.g. Mocca Chocolate).

5.1.6 Use of the Term Chocolate

Products not defined under this Standard, and where the chocolate taste is solely derived from non-fat cocoa solids, can carry the term "chocolate" in their designations in accordance with the provisions or customs applicable in the country in which the product is sold to the final consumer and this to designate other products which cannot be confused with those defined in this Standard.

5.2 Declaration of Minimum Cocoa Content

When required by the authority having jurisdiction, products described under Section 2.1 of this Standard, except for *white chocolat*, shall carry a declaration of cocoa solids. For the purpose of this declaration, the percentages declared shall be made on the chocolate part of the product after the deduction of the other permitted edible foodstuffs.

5.3 Labelling of Non-Retail Containers

5.3.1 Information required in Section 5.1 and 5.2 of this Standard and Section 4 of the *Codex General Standard for the Labelling of Prepackaged Foods* shall be given either on the container or in accompanying documents, except that the name of the product, lot identification, and the name and address of the manufacturer, packer, distributor and/or importer shall appear on the container.

5.3.2 However, lot identification, and the name and address of the manufacturer, packer, distributor and/or importer may be replaced by an identification mark provided that such a mark is clearly identifiable with the accompanying documents.

6 Methods of Analysis And Sampling

6.1 Determination of Centre And Coating of Filled Chocolate

All methods approved for the chocolate type used for the coating and those approved for the type of centre concerned.